Ancient Cosmologies

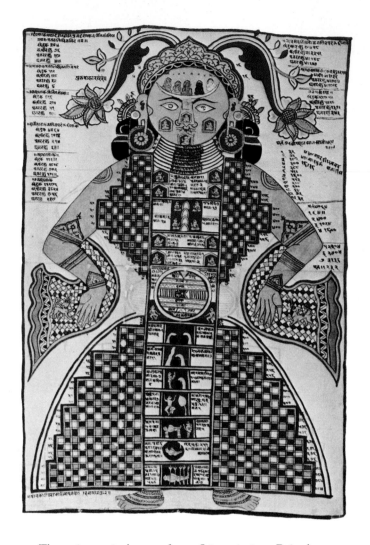

The universe in human form. Jain painting, Rajasthan (western India), ? eighteenth century. The human attributes are symbolic and decorative; for more purely diagrammatic representations see Plates 21 and 22; only the central column is inhabited by living beings, (from A. Mookerjee, *Tantra Art*).

Ancient Cosmologies

Edited by

CARMEN BLACKER AND MICHAEL LOEWE

With contributions by

J. M. PLUMLEY   W. G. LAMBERT
RABBI LOUIS JACOBS   JOSEPH NEEDHAM
R. F. GOMBRICH   EDITH JACHIMOWICZ
H. R. ELLIS DAVIDSON   G. E. R. LLOYD
PHILIP GRIERSON

London   George Allen & Unwin Ltd
Ruskin House   Museum Street

First published in 1975

ISBN 0 04 100038 2

Printed in Great Britain
in 12 point Fournier type
by Cox & Wyman Ltd
London, Fakenham and Reading

Acknowledgements

The editors and publishers wish to express their thanks for the kind help given by Dr Raymond Allchin and Dr Nicholas de Lange. They would also like to thank the following for their assistance in providing the illustrations for this book: the British Library; the British Museum; the East Asian History of Science Library and the University Library, Cambridge; the Manx Museum and National Trust; the Statens Historiska Museum, Stockholm; the Morgan Library, New York; the Topkapi Sarayi Müzesi, Istanbul; the Uppsala University Library, Sweden; Bavaria Verlag-Heinrich Frese, Munich; Cambridge University Press; Faber & Faber; Mr Ajit Mookerjee; George Wittenborn Inc.; Alex Poignant; Thames & Hudson; Paul Hamlyn; Harvard University Press; University of Chicago Press; McGraw-Hill; Eaton & Mains, New York; and Gotlands Fornsal, Visby.

Contents

Illustrations

Frontispiece

The universe in human form. Jain painting, Rajasthan (western India), ? eighteenth century. The human attributes are symbolic and decorative; for more purely diagrammatic representations see Plates 21 and 22; only the central column is inhabited by living beings.

Introduction

The nine chapters in this book are, with a single exception, based on lectures delivered in Cambridge University in 1972. All seek to answer the question: what was the shape of the universe imagined by those ancient peoples to whom all modern knowledge of geography and astronomy was inaccessible? How did the ancient Egyptians, Babylonians, Jews, Indians, Chinese, Arabs, Greeks and Norsemen conceive the form of the cosmos which accommodated not only the known face of the earth and the visible heavenly bodies, but also those other worlds which it was deemed necessary to locate comprehensibly in space – the realms of the dead, both blessed and damned, and the countries inhabited by gods and demons?

A remarkable range of answers presents itself. Professor Plumley's Egyptians lived on a flat island bisected by a great river, over which hung a heavenly canopy supported by four posts. Professor Lambert's Babylonians saw themselves not, as is commonly supposed, on the slopes of a vast ziggurat, but in a universe of two superimposed layers held together by a cosmic rope or staircase. Rabbi Jacobs, with the aid of Rabbinical commentary, explains how in the Old Testament universe the earth was related to the mysterious 'firmament', to the waters above the firmament, to the fountains of the abyss, to limbo and to the storehouse of winds.

For ancient India Dr Gombrich introduces us to several cosmologies – Vedic, Brahmanic, Jain and Buddhist. In all of these the universe is transfixed by a vast axial mountain, about which at varying levels are ranged the continents of our own world and the layers of heavens and hells demanded by the Indian doctrine of reincarnation. For the Hindus the universe was a round egg, covered with seven concentric shells composed of different elements. For the Jains, on the other hand, it was shaped rather like an enormous man, or sometimes an hour-glass with a narrow waist, and was measured in terms of 'a peculiar unit, the *rajju*, which is defined as the space covered by a god in six months if he flies at 2,057,152 *yojana* in a "blink"'. And for the Buddhists the universe consisted of three horizontal layers: the world of desire, in which lay our own earth, surmounted first by the world of form, and then, floating well above the summit of the axial mountain, by the mysterious world of no-form, which is clearly a translation into spatial terms of various mystical states of consciousness.

The powerful image of the *axis mundi* appears again in Dr Davidson's Norse universe, with its huge Tree, at the centre of the nine worlds, its roots, twined about with serpents, stretching down into the regions below, its branches reaching to the heavens.

As for the Chinese universe, Dr Needham chooses three theories for consideration from the large range of indigenous cosmological doctrine: the celestial sphere, the dome and the 'vision of infinite space with celestial bodies at rare intervals floating in it'. He considers also the interesting problem of the location of the world of the dead. Heavens and hells, he tells us, only appeared in the Chinese cosmos with the advent of Buddhism. From the earlier Chinese universe 'ethical polarity' in the destination of the dead was notably absent; just and unjust alike were relegated to an underworld called the Yellow Springs.

In Greece, Dr Lloyd tells us, no one orthodox and overriding cosmological theory held sway. A plurality of theories confronts us, and we read of philosophers arguing for the universe as a living organism with a soul, as an artefact, the product of a craftsman or design, or as a political entity. And, although geocentric theories with man firmly at the centre were preferred, we hear as early as the third century BC of theories based on heliocentricity, with a daily axial rotation of the earth and an annual revolution of the earth round the sun.

Dr Jachimowicz's Arabs also give us a mixed tradition. On the one hand we read in the Koran and the *mi'rādjnāma* literature of Mahomet making his ascent to the Throne of God through ten different cosmic levels. But we also find the medieval Arab philosophers writing of a cosmos richly influenced by Aristotelian and Ptolemaic thought, a universe of nine spheres, ranging from the sub-lunary earth to that of the Sphere of Spheres above the fixed stars.

In some cosmologies space is inseparable from time, and no account of the shape of the universe makes sense unless we also know how it came to be so in the first place. For some ancient peoples, therefore, the account of the creation of the universe is an essential feature of their cosmology. Rabbi Jacobs shows from both Biblical and Rabbinical sources how much this question worried the Jews. From Dr Needham, on the other hand, we gather that the notion of creation was not of prime importance to the Chinese. The Greeks and the Jains were likewise uninterested in beginnings. For the Indians time was conceived on the same bafflingly vast scale as space, and Dr Gombrich recounts the system of enormous cycles of time, found in both Puranic and Buddhist texts, through which the universe must pass for all eternity. For the Greeks, Norse, and Chinese cosmic systems, too, time was cyclical. Indeed the notion of linear time is only found expressed with any clarity in the Old Testament universe.

The shape of the universe also to some extent depends on its component physical elements. To the question, of what is the universe composed, we again find a remarkable range of answers. From the relatively simple earth, air and water of the ancient Egyptians we are introduced to the sophisticated theories propounded by the Indians, Greeks and Greek-influenced Arab philosophers in which four, five or six elements are subject to mutual influences and combinations.

Finally from Professor Grierson we have a discussion of the heritage received by Europe from these ancient systems, a double inheritance combining Greek ideas of the structure of the universe with Jewish notions as to its origins. The combination was not always a coherent one. There were contradictions to be discovered between the Ptolemaic system received from Greece, with its crystalline and translucent spheres, and the Biblical account of the Creation, Fall and Redemption of man. Throughout the Middle Ages, however, such a system, increasingly elaborated along lines of astrological correspondence, was the world-view of most of Europe. Its breakdown, which Professor Grierson attributes not only to Copernicus, Kepler and Darwin but also to the discovery of America, the Portuguese voyages to India and the habit of looking to experience rather than to ancient authority for the matrix of knowledge, did not come until after the Renaissance.

It should be explained that the system of dating used throughout the book is not entirely uniform. Rabbi Jacobs understandably prefers the Jewish convention, while Dr Needham's pluses and minuses, already familiar to readers of his celebrated work on Chinese science, are his own attempt to make our Western system more acceptable to non-Christians. Likewise Dr Needham's own is the system of romanisation into which he has transcribed his Chinese names.

Cambridge 1973   C. E. B.

I

The Cosmology of Ancient Egypt

J. M. PLUMLEY

Professor of Egyptology, University of Cambridge

Most English dictionaries define the word 'cosmology' as a metaphysical view of the universe as an ordered whole. Reasonable as this definition may seem to us, it must be recognised that it is a definition which is only meaningful to an age which has inherited an independence of thought which finds no difficulty in accepting a dichotomy between what might be called Church and State, clerical and lay, religion and science. It is therefore important that when we speak of what the Ancient Egyptians meant by cosmology, we should not forget that this same people belonged to the pre-Greek era, and that for them neither the content nor the language of what we know as philosophy existed.

Nineteen centuries divide us from the Ancient Egyptians. Their expressions and processes of thought sprang out of beliefs now long lost. Even if it were possible to recover some of those ancient beliefs in their entirety, they would be utterly foreign to us in our modern European setting. And furthermore, because we live in a different age, we must not expect the Ancient Egyptians to have asked some of the questions which we ask about the world in which we find ourselves.

In addition we must recognise that there are problems of satisfactory transmission of ideas, for we are trying to understand a language which has ceased to be spoken for hundreds of years, and which, over three millennia, revealed in its written

B

forms many linguistic changes and developments. It would be true to say that, while we can usually grasp the general meaning of what is recorded in one or other of the forms of hieroglyphic writing, we cannot be certain of the exact meaning of certain passages and phrases. Nuances will, for the most part, forever escape us. Not only do we lack knowledge in both the grammatical and lexicographical aspects of the language of Ancient Egypt, but many of the similes and metaphors are meaningless to us, since we no longer live in the kind of society that provded such similes and metaphors, and we are no longer sufficiently acquainted with the beliefs that were once part and parcel of the life of the Nile Valley.

For many centuries the Ancient Egyptians lived comparatively undisturbed in their river valley; for the desert region of the Sinai Peninsula, while negotiable by small, experienced and determined bands of traders from the other civilized parts of the Near East, presented a formidable natural barrier against large-scale foreign invasion. During this long period of peace the ideas and mental conceptions of the Ancient Egyptians continued to develop along their own peculiar, insular lines of thought, to be represented repeatedly in pictures and writing alongside new and often contradictory mental images. Indeed it is this abundant wealth of illustration which, while being a source of information, has frequently proved an embarrassment to modern students of Egyptology. The plain fact is that the extent of the available material is such that one is forced to concur with an eminent German Egyptologist, Adolf Erman, who wrote nearly seventy years ago: the material 'is so endless and impossible to grasp . . . it is in fact too great'.

Or, to use an analogy, it is as if one were to find oneself in a vast museum housing a collection of materials which would seem to have been brought together on the principle that the collection must be all-embracing, even at the expense of being scientifically selective. Indeed it is not so much a museum as a gigantic depository. On closer examination one would find that, while the greater part of the exhibits had labels, many of the labels were hard to read. In some instances this would be because the labels had faded with age. It would soon become clear that in other instances later hands had attempted to re-write the labels or to add further details and explanations. Not infrequently these later additions would only serve to add to the confusion. It would be found that some of the exhibits had no labels at all, simply because their meaning and purpose was obvious. But what was once obvious to the ancient curators is not necessarily so to us. Sometimes there would be no label, because even to the Ancient Egyptians the object was so old that they themselves had lost all knowledge of its name and original purpose.

Though for the reasons stated we cannot understand fully the thoughts of the Ancient Egyptians, we can still put ourselves into the same physical environment and geographical setting which so greatly shaped their ideas. It is of course a fact that the Nile Valley has undergone various changes since the time of the Pharaohs. The most significant of these changes has been the complete cessation of the inundation or annual rise of the Nile waters as the result of the recent construction of the Aswan dams. Nevertheless, most of the climatic and geographical features of Egypt are the same today as they were in ancient times.

Contrary to modern usage the Ancient Egyptians orientated themselves to face southwards. At their backs lay the Mediter-ranean and the rest of the ancient world. The west was for them the right, and the east the left. Immediately before them lay the plain of the Delta, 112 miles long and 181 miles in width with great lakes on either side – Lakes Mareotis, Burlus, and Menzaleh, etc. In ancient times this alluvial plain was divided by three great branches of the Nile and interlaced with many waterways both natural and artificial. Near the locality where the Nile divided, the eastern and western deserts closed in on the river valley to create a narrow corridor of cultivable land, rarely exceeding eighteen miles in width at any point, and extending southwards for 625 miles as far as the First Cataract at Aswan. Here a natural river barrier of granite out-cropping formed the historic boundary of Egypt, being known in ancient times as 'The Door of the South'. Beyond it lay the lands of Nubia and the Sudan which in the course of time were to come under the direct rule of Egypt.

The striking contrast in colour between the cultivated banks of the river valley and the bordering deserts, most easily seen today from the air, did not go unnoticed by the Ancient Egyptians, who distinguished the two by describing them as 'the Black Land and the Red Land'. It was the Black Land, Keme, which was Egypt proper. Every other place, not in the Valley of the Nile, whether the limestone cliffs which bordered the Valley or the distant countries of Nubia or Syria, all were lumped together in the eyes of the Ancient Egyptians as 'mountainous country', while the hieroglyphic writing of the name of a foreign country was usually accompanied by the determinative sign representing a line of hills or mountains. The Ancient Egyptian was always aware that in order to leave Egypt one literally had to walk up out of the Valley of the Nile and into the bordering hills.

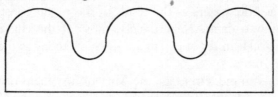

1. Hieroglyph denoting a foreign country.

Having little, if any, knowledge of other lands than their own, it is not surprising that the makers of the most ancient Egyptian picture of the world should have been inspired solely by the physical and geographical conditions of the Nile Valley. Accordingly they thought of the world as being a bank of earth divided in the middle by the Nile and surrounded by water, the Great Circuit or the Great Circular Ocean. This mass of water was itself the product of the first of the gods, Nun, from whom the world came forth, and in whom the Nile and the rain had their origin.

Above the earth was the expanse of the sky, separated from the earth by the air, and held aloft like a great flat plate by four supports at the corners of the earth*. In some representations these supports are shown as poles or forked branches, such as might be used for holding up the corners of an awning. In other instances they were spoken of as being four great mountains. It is clear that although the conception was first applied to a small local universe, as the knowledge about other countries grew, the confines of the world picture were accordingly extended in all directions, but apart from these extensions and the varying of a few details, the overall conception of the world was not changed significantly.

At times the sky appeared to the Ancient Egyptians to be not unlike a gigantic canopy stretched out over the valley. On the other hand its colour, unbroken for the greater part of the year by clouds, suggested to them that it might be some kind of celestial ocean across which the sun sailed in a boat by day and the moon by night. During the hours of darkness the same sky was lit by myriad points of light – the stars – the appearance of which in the clear night air of Egypt was no less a pageant of splendour to the gaze of the most ancient dwellers beside the Nile as it is to modern man.

It was a daily event that the sun rose in the east and, passing high over the earth, descended at evening into the western hills. Apart from an annual shift in its path towards the south during the winter and a return to the north in the spring, there appeared to be no other variation in its daily journey. When it disappeared in the west, not only did darkness envelop the land, but the cold air from the rapidly cooling deserts descended into the valley of the Nile. Not illogically the earliest inhabitants of Egypt concluded that without the sun's light and warmth death would engulf the world. The eagerness with which they welcomed the sun's return at dawn is attested in the various versions of hymns to the sun which have survived from all periods of the history of Ancient Egypt. But they were at a loss to explain the sun's re-appearance in the east other than by postulating that during the hours of the night it had travelled beneath the earth upon which they lived. Further,

* See plate 1.

it seemed to them that the sun travelled on through another world, possibly a duplicate of the world in which men lived, but a world which lay deep underground. We may smile at so naïve a conception, but not so many centuries have passed since our own forebears thought of heaven above and hell below and 'merrie middle earth' in between.

As to how the sun made its daily progress across the face of the sky there were a number of explanations. In some the sun was the eye of the sky, beaming down upon the earth. In others the sun sprang like a glittering hawk from the eastern hills to soar triumphantly into the vault of heaven. In yet other explanations its journey across the heavens was in a splendid celestial boat*. Born at dawn as the child of the sky goddess it entered the solar bark to mount the heavenly ocean, rapidly growing into full manhood as the point of midday was reached, and then sinking down as an ageing old man into the west. In some accounts it was born as a calf from the womb of the sky then thought of as a great celestial cow. In yet another account, which employed a very earthy simile, it was considered to be a gigantic beetle rolling the fiery ball of the sun before it, just as the common scarabaeus beetle rolls its ball of dung before it into its lair to devour it in the darkness.

The appearances and movements of the moon in the night sky and its occasional presence in the day undoubtedly posed problems for the early inhabitants of Egypt. In the first place the moon possessed less light and gave no warmth. Not only were its appearances in the sky less predictable, but its disc altered in size and shape. Its seemingly erratic movement, its growth and decrease, suggested that like the sun it was a living entity, but unlike the sun it was less dependable and less powerful. From the earliest times various explanations were advanced for its apparent irregularity and change. At Thebes it was worshipped under the name of Khonsu. The root meaning of the name Khonsu means 'to travel' and more especially 'to travel through a marsh'. It is not difficult to see that the progress of the moon was not unlike a man travelling through

* See plate 4.

a marsh between the clumps of reeds and waterplants, so that his progress is not always easy to perceive or anticipate. And there is no doubt that at a very early period the moon became the object of careful observations, observations which led to its use as a means of time reckoning in relation to the seasons. As in other parts of the Ancient East the creation of a Lunar Calendar must have been amongst the earliest achievements of man.

To the most observant watchers of the night sky in Egypt it was apparent that there was movement in the stars. What at first sight seemed to be a fixed and invariable pattern of points of light or celestial lamps was seen to move slowly across the expanse of heaven during the hours of darkness. Continued observance over the seasons revealed that some of the fixed patterns of stars dropped below the horizon to remain invisible for considerable periods of the year before returning once more in the night sky. In the observation of this annual movement, the appearance of one bright star, Sirius, the so-called Dog Star, was noted by the Ancient Egyptians in reference to the commencement of the annual inundation. This was a most important event in the life of an agricultural people, and incidentally for the Egyptologist too, for whom the few written records of the rising of Sirius are a most valuable aid in establishing a more exact chronology. The slow but demonstrable movement of the constellations and the more obvious movements of the five planets visible to the human eye were seen as indications of some kind of life in the celestial regions, and it is not difficult to see that the conclusion followed that the heavenly bodies were in fact living entities. Notwithstanding the influences of such conceptions the Ancient Egyptians, though never the equals of the Babylonians as astronomers, possessed a knowledge of the heavens which was far from negligible. Charts of the heavens, carved or painted on the ceilings of tombs and temples*, dated tables referring to the movements of the stars, some astronomical treatises, notes in religious writings, their division of time and their calendar show quite clearly the attention which the Ancient Egyptians

* See plate 2.

gave to celestial movement, even if their explanations appear to us to be utterly unscientific.

In trying to understand the thought of the Ancient Egyptians a number of facts must be borne in mind. For them the whole universe was a living unity. Even those parts of the physical world which we are accustomed to think of as inanimate, e.g. stones, minerals, water, fire, air, etc., partook of a common life in which men and women and animals and birds and fishes and insects and plants and even the gods themselves shared. The Ancient Egyptian, in common with other ancient peoples, could not conceive of anything that was not alive in some degree. Having no neuter gender in his language it was necessary for him to classify this or that thing as masculine or feminine. By making such distinctions he did not imply that there was any manifestation of what might be termed physical sexual differences. Rather he was concerned with representing qualities, characteristics, tendencies as exhibited by living creatures irrespective of what their physical sexual distinction might be. This being so, it is not surprising that in his attempts to express abstract ideas he employed personification, and it is this use of personification which in no small measure accounts for the multitude of deities, both male and female, which is so marked a feature of Ancient Egyptian religion.

It must also be noted that while the Egyptian thought of the world as a living unity, he was aware that much of it was subject to decay as well as to growth, that there was both a coming into being and a passing out of being, that there was birth and there was death. On the other hand he also perceived that there were some things which appeared to be indestructible, permanent, never changing. In particular the seemingly unchanging and ever-continuing physical conditions in the Nile Valley fostered in his mind a sense of determined continuity in creation. Ever before him were the daily rise and setting of the sun, the annual rise and fall of the waters of the Nile, the yearly recurring cycle of life. Added to these was the equable and rarely changing climate. It is not surprising that the monotonous regularity of life in the Nile Valley tended to produce the sense of 'as it was in the beginning, is, and ever shall be'. Perhaps it is not an exaggeration to say that the Egyptian mind was pervaded by an eternal now. Indeed, something like a sense of timelessness is latent in the ancient language. Throughout the long history of Egypt the conception of the timeless state was never absent, be it the Old Perfective form of the verb in Ancient Egyptian or the Qualitative form in Coptic, and it is not without significance that while it was always possible to negate verbal forms which denoted action, negation was not possible, at least not until much later and then only rarely, with those verbal forms which denoted state.

Though the foregoing remarks are a reminder that the men of Ancient Egypt used thought patterns very different from ours, they did not thereby differ from us in asking certain questions which we find ourselves asking. As has been mentioned earlier, we should not forget that because of the limitations of their knowledge there are some questions which we must not expect them to have asked. But at least the men of those ancient times would have been at one with us in asking how the world came into being, how it all began.

The actual asking of the question indicates that the Ancient Egyptians could speculate that there had been a time when the world as they knew it did not exist, though it was difficult for them to imagine that there could ever have been a time when there did not exist something, however nebulous, out of which the world itself was created. For them that something was a state described as the Primeval Waters. Every Egyptian Creation Story, of which there are three major accounts and the remnants of an unspecified number of others, starts with the basic assumption that before the beginning of things there was a Primordial Abyss of Waters, everywhere, endless, and without bounds or direction. This was unlike any sea which has a surface, for here there was neither up nor down, no distinction of side, only a limitless deep – endless, dark, and infinite. As one Spell from the Coffin Texts expresses it, 'in the infinite, the nothingness, the nowhere, and the dark'. This watery state was personified as the god Nun, who is sometimes called in the texts 'father of the gods'. He remained, however, a purely intellectual concept, possessing neither temple or worshippers. Not surprisingly, representations of Nun are rare. In one papyrus of the Book of the Dead he is depicted as a male figure plunged up to his waist in water, elevating his arms to support the boat of the sun*. In the boat the solar disc is held aloft by the scarab beetle which is flanked on each side by the deities who, it was believed, came forth from Nun. The name of this most ancient of gods passed on into Coptic, the last stage of the old language used in Christian Egypt. Then, as *Noun*, it came to mean the deep abyss of hell or the unfathomable depths of the sea.

The basic principle of Egyptian cosmology may therefore be said to be the Primeval Waters, which existed before the beginning and which would last for ever. Whereas we think of our world as a speck of matter moving through boundless space, the Ancient Egyptians thought of their world as a cavity, rather like an air bubble, floating in the midst of the limitless expanse of Nun, the Primeval Waters. It seemed logical to them to believe that those waters were to be met everywhere at the limits of their world, and beneath the earth and above the sky. Just as for the Hebrews, so for the Egyptians there was a firmament dividing the waters above from the waters below. Likewise, it was thought that the seas, the rivers, the rain from heaven, and the waters in the wells, and the torrents of the floods were parts of the Primeval Waters which enveloped the world on every side. Though there is a hint in some texts that the Primeval Waters might one day like Noah's Flood come seeping in as 'the windows of heaven were opened and the fountains of the great deeps broken up' to engulf the world, by and large Nun, the personification of the waters, was thought to be a beneficent god who guarded and kept in check the demonic powers of chaos, represented by gigantic dragon-snakes.

The three most important and best attested cosmogonies are those associated with the religious centres of Hermopolis, Heliopolis, and Memphis.

* See plate 3.

At Hermopolis, so called by the Greeks because it seemed to them that their deity Hermes most closely represented the Egyptian god Thoth, the four characteristics of the Primeval Waters – depth, endlessness, darkness and invisibility – were personified by eight beings*. That is, each of the four characteristics was represented by a male and a female form. Thus there were Nau and Naunet (depth), Huh and Hauhet (endlessness), Kuk and Kakwet (darkness), and Amun and Amaunet (Invisibility). The number of these deities gave rise to the native name of Hermopolis, Khnum (in Coptic *Shmoun*) literally 'eight' or 'Eight Town'. The deities were worshipped there as genii, the males being represented with the heads of frogs and the females with the heads of serpents. It was believed that the eight swam together and formed the primeval egg in the darkness of Nun. According to one version, out of this egg burst the bird of light. But according to another version, the egg contained not light but air. In yet another version, which was associated with the district of Thebes, the egg was laid by a goose†, the Great Primeval Spirit called *Kenken Wer*, which means literally 'the Great Cackler' whose 'voice broke the silence . . . while the world was still flooded in silence'.

The account is complicated by the presence of the supreme god of Hermopolis, Thoth‡, who was considered to be the head of the eight genii. It is possible that in early times Thoth had been a creator-god in his own right, but in the dynastic period this function had become obscured, for he was henceforth credited with the invention of hieroglyphic writing and was regarded as the original lawgiver (*štn hpw*), the repository of all learning, both sacred and profane, and the master of *Hike*, which may be rendered loosely as 'magic'.

In a variant of the Hermopolitan cosmogony a lotus flower was said to have arisen in the beginning out of the primeval waters. When its petals, which had been closed in the primeval darkness, opened, the creator of the world in the form of a beautiful child sprang from the heart of the lotus§. This

---

* See plate 10.   ‡ See plate 8.
† See plate 5.   § See plate 11.

was the infant sun, who immediately spread his rays of light throughout the world. Henceforth each day at evening, when the light fades, the lotus closes its petals around the sun and shelters it during the hours of darkness only to release it again when the next morning dawns.

The Hermopolitan creation account is further confused because it would seem that during the First Intermediate Period in the history of Egypt (c. 2200–2000 BC) the cosmogony of Hermopolis became mingled with that of Heliopolis, with the result that many of the details in the cosmogony of Hermopolis were ousted from the creation account, current in Heliopolis.

At Heliopolis the Creator-god who held pride of place was Atum, whose name probably meant 'the Complete One'. According to the earliest versions of this cosmogony Atum emerged out of the primeval waters in the form of a hill. A later recension of the story states that Atum arose out of the waters seated or standing upon the hill. Some texts speak of Atum as the child of Nun, others affirm that he was self-created. In his emerging like a hill Atum would have resembled the life-engendering hillocks which appeared regularly in the river when the waters of the inundation began to recede. The Egyptians could not fail to notice that every year, shortly after each little hill broke through the surface of the river, its top would begin to swarm with life.

The Primeval Hill seems to have had no fixed form, and there are various representations of it. The early formalizing of the Hill into an eminence with sloping sides or a platform surrounded by steps is probably what the Step Pyramid at Sakkara, and possibly the later pyramids with unbroken sloping sides, were supposed to represent. Indeed it has been suggested that this formalizing of the Primeval Hill may be the prototype of the later obelisk which was associated with the worship of the sun, of whom Atum was considered to be a form. Not unnaturally the priesthood of Heliopolis claimed that it was at Heliopolis that Atum emerged in the beginning. But the priesthoods of other religious centres were equally ardent in claiming that it was in their vicinity that the place of creation was to be

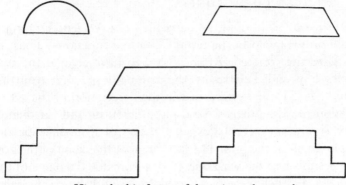

2. Hieroglyphic forms of the primeval mound.

found. Thus at Hermopolis the site of the beginning was a rectangular space surrounded by a wall; within lay a pool of water named the 'Lake of the Two Knives', symbolising Nun. In the middle of the pool was an island upon which rose a hill from which it was believed light originated. This hill was a place of pilgrimage and the setting for the performance of much ritual. And if further proof were needed to confirm the veracity of the claim, pilgrims were shown the remains of the primeval egg from which the creator of the world came forth. Other religious centres, undeterred by such evidence, claimed for their localities the actual site of creation. At Memphis the whole district round the city was named *Ta-tjenen*, i.e. 'the land rising', and at Thebes it was asserted that the Primeval Hill was situated in the temple area of Medinet Habu, on the West bank of the Nile.

Spell or Utterance 600 of the Pyramid Texts (*c.* 2400 BC), the oldest surviving corpus of religious literature from Egypt, reads: 'O Atum-Khoprer, thou becamest high in the height. Thou didst arise up as the *benben* stone in the Mansion (i.e. Temple) of the Bennu Bird (i.e. the Phoenix) in Heliopolis. Thou didst spit out Shu. Thou didst expectorate Tefenet, and thou didst set thy arms about them as the arms of a *ka*-symbol that thy essence might be in them.'

This passage refers to the time when Atum was alone in the Cosmos, when no world had been created, and when he contained in himself everything that was to come into being. In order that the world might come into being, Atum had to create it out of himself, an act which the Ancient Egyptians could only explain in the terms of natural functions. Thus, in the Utterance quoted, Atum is said to have created the two deities Shu and Tefenet by the action of spewing from his mouth. It is to be noticed that in the description of the act of creation the Egyptians were using a favourite and for them a very significant verbal device, the use of assonance or the employment of the pun. The facts are that the name of the god Shu is similar to the verb *ishesh* 'to spit', while the name of the goddess Tefenet resembles the word *tef* with much the same meaning as *ishesh*. For the Ancient Egyptians the name of a person or a thing was potential of actual being, and to utter the name of something was to bring what was named into being. In the cosmogony of Memphis this conception of the power of speech was to have far reaching consequences in the account of the creation of the world. It was this kind of reasoning which led the people of the Ancient World to be circum-spect in their mentioning of certain persons and things, and caused them to resort to the employment of euphemisms when some particularly dangerous person or thing had to be referred to in speech.

In another version of the Hermopolitan cosmogony Atum, having no consort, achieves his purpose of creation by mating with his own hand. This account, which may seem indecent to Western ears, stems from a primitive naturalistic view of the world which could only account for creation in the terms of physical generation. But for the Ancient Egyptians both ver-sions would have been seen as complementary and not alterna-tive or contradictory.

The two children of Atum, Shu and Tefenet, were them-selves personifications. Shu was the air and Tefenet, his sister, appears to have been associated with moisture, and is therefore the goddess of damp, mist, dew and rain. To Shu and Tefenet were born the two deities, Geb and Nut. Geb was the earth god, and Nut was the sky goddess. In the beginning both were locked together in a close embrace, but their father Shu separ-ated them, lifting Nut high above himself to form the arch of heaven, and leaving his son Geb lying prone upon his back to become the earth. Shu then remained in between them to be the air between heaven and earth*. In the numerous representations of this story the goddess Nut appears, her body painted blue and encrusted with stars, standing and bending over the reclining body of her brother, her hands touching the ground. Geb is usually coloured green to represent the vegetation. Sometimes the figure of Nut as a woman is replaced by that of a cow. Sometimes she was called the mother of the sun, for as the sky goddess she was supposed to swallow the sun at evening, and to give birth to him when he came forth from her womb on the following morning†. According to one account Nut married her brother Geb secretly and against the wish of the sun-god Re. In his anger Re had the couple forcibly separated by Shu, and further decreed that Nut should never be able to bear a child in any given month of the year. According to another tradition, preserved by Plutarch, the god Thoth took pity on Nut. By playing a series of games of draughts with the moon he managed to win a seventy-second part of the moon's light, out of which he created five extra days. Since these days did not belong to the official Egyptian Calendar of twelve months of thirty days, totalling 360 days, Nut was able on the five days before the New Year began to give birth to five children, Osiris, Horus, Set, Isis, and Nephthys, all of whom became important deities in the Egyptian pantheon.

A papyrus in the British Museum (B.M. Pap. 10188), *The Book of Knowing how Re Came into Being, and of Overthrowing Apepi*, contains two versions of the Hermopolitan cosmogony. The first of these is interesting because it contains a reference to the creation of Mankind. Here the creator-god is named Neb-er-Djer, i.e. 'the Universal Lord'. Neb-er-Djer is quoted as saying: 'Now after the creation of Shu and Tefenet I gathered together my limbs. I shed tears upon them. Mankind arose from the tears which came forth from my eye.' In a fragment which has been preserved in a magical work of the First Intermediate Period called *The Book of the Two Ways* the Universal

* See plate 9.   † See plate 6.

Lord says: 'The gods I created from my sweat, but mankind is from the tears of my eye.' In these two extracts there are allusions to two creation stories. The gods, on the one hand, are the exhalations of the god himself, whereas mankind comes from the tears of his eye. This essential difference between the gods and man is nowhere else mentioned. In both accounts the origin of mankind is linked with the assonance between the Egyptian words for 'tears' and 'mankind' – *remeyet* and *romet*. Effectively isolated from the rest of the Ancient World of many generations in the earliest periods of their history, the Egyptians tended to consider themselves as being the only true men, *romet*, the favoured of the gods, and in their earlier history the term *romet* was reserved exclusively for the men of Egypt and was not applied to foreigners. Indeed, through the whole of their history they were slow to welcome men of other nations into their midst.

At some time during the Old Kingdom (*c.* 2700–2200 BC), when Memphis was the capital city of Egypt, it seems that a need was felt to attempt a reconciliation between the cosmogony of Heliopolis, in which Atum was the creator-god, and the cosmogony of Memphis, which attributed the act of creation to Ptah*. Not surprisingly, the priesthood of Memphis were not prepared to allow their deity Ptah any place less than first in this attempted reconciliation of differing beliefs. Nor does it seem that they were content to countenance what had been the practice elsewhere, a synthesis of beliefs and a creating of composite deities. Instead, they conceived an approach so novel as to be reckoned one of the great religious achievements of early man. For the details of the Memphite cosmogony we are dependent on a hieroglyphic text engraved upon a slab of black basalt now in the British Museum, which was presented to the Museum in 1805 by Earl Spencer, who had obtained it from the ruins of the temple of Ptah in Memphis. It had once stood upright in the Temple, but in the course of the years it had been thrown down so that large portions of the text had been obliterated by the passage of many feet across it and by

* See plate 7.

its having been used as the nether stone of a mill. Not only were the signs barely legible in places, but when the text was first examined it revealed an unusual arrangement of signs, and for many years it defied all attempts by scholars to read it. It was not until the end of the nineteenth century that the American Egyptologist, James Breasted, recognised that the lines of inscription were not to be read in the usual manner with the hieroglyphic signs facing the beginning of the sentences, but in the opposite direction. His pioneer translation of the text encouraged the eminent German Egyptologist, Adolf Erman, to re-examine the inscription afresh and to produce a skilful analysis of its contents. Finally in 1928, over a century after its discovery, Erman's distinguished pupil and successor, Kurt Sethe, published what can be regarded as the final edition of the hieroglyphic text with a translation, comprehensive notes and a most instructive introduction.

It had already been confirmed by both Breasted and Erman that the text was a copy of a much older examplar. During the eighth century BC the existence of a very ancient worm-eaten wooden archetype in the temple of Ptah at Memphis had been brought to the notice of the reigning Pharaoh, the Ethiopian Shabaka. Although the inscription was badly eaten away in places, the portions of the text which remained were considered by Shabaka to be of such importance that he gave orders for a copy to be made in hard stone so that their contents might be preserved for future generations. Sethe, in his study of the inscription, was able to show that the text was the remains of the libretto of a sacred drama or play which was probably performed during the annual celebration of certain of the great religious festivals at Memphis. During the course of the unfolding of the action of the drama various gods and goddesses made speeches, here and there explanatory remarks were added by unnamed beings, and the fundamental beliefs of the theology of Memphis were laid bare. From the text it emerges that during the Old Kingdom (c. 2700–2200 BC), one priesthood at least had been able to formulate the doctrine that the god who had created himself and heaven and earth, who was the maker

c

of gods and men and animals and all that exists, was a Spirit and the Eternal Mind of the Universe. As in the cosmogonies of Hermopolis and Heliopolis we find mention of the primeval god of the Waters, Nun and a female counterpart Nunet. But in the Memphite cosmogony these are said to be the products of the Eternal Mind, Ptah, who manifests himself in many ways and under many aspects. The ancient gods of the other cosmogonies, including Atum, are said to be contained in Ptah. 'They have their forms in Ptah' and are nothing but Ptah. Atum is stated to be the heart and tongue of Ptah, and the divine forms of these are the gods Horus and Thoth.

To quote from the text: 'Creation took place through the heart and the tongue as an image of Atum. But greatest is Ptah, who supplied all gods and their faculties (*kas*) with (life) through this heart and tongue – the heart and tongue through which Horus and Thoth took origin as Ptah. His *Ennead* (of gods) are in front of him as teeth and lips which are nothing else but the seed and hands of Atum, for the Ennead of Atum was born of his seed and his fingers, but the Ennead are the teeth and the lips in his mouth which has pronounced the names of all things, from which Shu and Tefenet have come forth, and which has fashioned the Ennead.'

Or in another passage: 'In this way were fashioned all the gods, Atum and his Ennead were complete. For every divine word sprang from what the heart had pondered over, and the heart had ordained. And in this way the faculties (*kas*) were made and their female counterparts (*hemset*) were determined, which make all food and all supply through this word. And in this way that man was declared just who does what is loved, and that man was declared wrong who does what is hated. So also life is given to the peaceable and death to the wrongdoer.'

In the Memphite cosmogony the creation of the world is presented in a highly intellectual manner as the combined action of conceiving through the intelligence (the seat of which is the heart) and creating through the spoken word or command (the seat of which is the tongue). The other gods are only this heart and tongue and lips and teeth of Ptah.

Interspersed in the account of creation by Ptah are passages which give a surprisingly accurate knowledge of physiological phenomena. Thus: 'for so it is that the heart and the tongue have power over all members, considering that (the heart) is in every body and (the tongue) is in every mouth of all gods, all men, all cattle, all worms, and all living beings, the heart conceiving thoughts freely and the tongue commanding freely. The seeing of the eyes, the hearing of the ears, and the breathing of the nose make report to the heart. And it is the heart which produces all cogitation, and it is the tongue which repeats all that has been thought out by the heart.' 'And so is produced all work and all craft. The activity of the arms, the walking of the feet, the movement of all the members, according to the order which the heart has thought out, have come forth by the tongue, and are put into effect in order to accomplish all things.'

To quote once more from the text: 'He is Tatjenen (the primeval hill at Memphis) who produced the gods from whom everything has come, whether food, divine sustenance, or any other good thing. So it has been found and understood that his power is greater than that of any other gods. And thus was Ptah satisfied after he had made all things and every divine word.'

The foregoing outlines of the cosmogonies of the Ancient Egyptians are an attempt to illustrate some of the ways in which they tried to answer the question as to how the world came into being. We look in vain in the great mass of religious texts for answers to some of the other questions which the fact of the existence of the Universe raises. If we want to know what answers the Ancient Egyptians preferred to such questions as 'Why was the World created, and to what purpose?' we must resort to what remains to us of the literature which has been given the name 'The Wisdom Literature'. Didactic treatises containing wise maxims and proverbial truths were very much to the taste of the Ancient Egyptians, and there is little doubt that some of these were very ancient. Much of the teaching in such works was based upon experience and had as its aim the way of life which a man ought to adopt in order to be happy. Much of the teaching was concerned with the observance of traditional etiquette and practical morality, but now and again, especially after the period of social upheaval and political disruption which followed the end of the Old Kingdom (c. 2200 BC), the writers of didactic treatises began to develop humanistic ideas and to ask why such and such an event came to pass. It is in these treatises that men raised the question why was man created, and even dared to postulate that man was the centre, even the purpose of creation, and that the world was his tool. Thus in a book of advice written by an unnamed king of the first Intermediate Period (c. 2200–2000 BC) for his son, *The Instruction for King Men-ka-re* we read: 'Well favoured are men, the flock of God. He made heaven and earth according to their desire. For them he drove back the monster of the primeval waters. He made the breath of life for their nostrils. They are his own image and made of his body. For them he shines in the heaven. In the same way he has made for them the plants, the animals, the fowls, and the fish to be their food.'

But whatever wise men might think about the purpose of creation and whatever might be the official doctrines about the way in which the creation came into being, there was the universal belief that what had been achieved in the beginning of time must be maintained. For mortal men the most essential task of earthly life was to ensure that the fabric of the Universe was sustained. The ancient cosmogonies were in agreement that obscure forces of chaos had existed before the world was created, and that, although in the act of creation they had been cast away to the outer edges of the world, they nevertheless continued to threaten to encroach into the world. The possibility of such a catastrophe could only be averted by the actions of gods and men, both working together to maintain the world order. That order which embraced the notions of an equilibrium of the Universe, the harmonious co-existence of all its elements and its essential cohesion for the maintenance of all created forms was summed up in the word Ma'at. In reality a pure abstraction, Ma'at was personified as a goddess and appears as a woman standing or sitting and wearing on her head a single ostrich feather, the hieroglyphic sign of her name*. Ma'at personified the cosmic order, but as a secondary development of this basic idea she also represented the ethics which recognise the need to act in all circumstances for the upholding of the universal order. Thus in the affairs of men she represented the two concepts, truth and justice. As the symbol of truth her figure or her feather was placed in one pan of the balance used for the weighing of the heart of a dead man in the judgement before Osiris the god of the dead. Her image was also worn as part of the insignia of the Vizier who was the supreme head of all the courts of justice and in his office as such was called 'the Priest of Ma'at' who 'spoke according to Ma'at and did not lie'.

Each day without fail creation must be renewed. This renewal was the work of the gods and their representative *par excellence* on earth, the reigning Pharaoh. And it was primarily for this work of renewal that the temples existed. While it was accepted that in the first instance the precarious maintenance of creation was only achieved through the ceaseless efforts of the gods, the gods in their turn depended upon the assistance of men, for they lived upon the earth in their mansions, the temples. The function of the temples was to provide for the gods security against the attacks of hostile forces, to nourish them with offerings of food and drink, to keep them clothed and in perfect condition for their divine tasks, and to ward off from them any influence which could impede their beneficent work. We are not to think, therefore, of the temples of Ancient Egypt as houses of prayer to which men might resort to seek comfort for their souls, but rather as government-supported religious establishments, restricted of access, and administered by staffs specially picked and trained. In theory the supreme officiant and head of each temple was the reigning Pharaoh, but for obvious reasons his office and duties were delegated to a high-ranking priest. And while every day a multitude of

* See plate 12.

different offerings was made to the various presiding deities of each temple, one offering in particular was never omitted – the culminating offering of the figure of Ma'at, for this was the supreme offering in that every other offering was virtually included in her.

However, the Ancient Egyptians were never lulled into a sense of false security. At all times in their history they were aware that the gods might be in heaven, but all was not thereby well with the world. Ma'at might indeed be the ruling principle of the universe, but without the exercise of some special power the threat of this or that adversary was very real. This dynamic power which had come into being on the day of creation was known as *Hike*. Though this term may be translated loosely as 'magic' it is important that we do not thereby imply notions which were foreign to the Ancient Egyptians. For them *Hike* was the power to be used for the protection of human beings and sometimes of gods. Its primary purpose was defensive. Thus, in the *Instruction for King Men-ka-re* it is said, 'God has given *Hike* to men as a weapon against adversity.'

Basically, *Hike* was founded on the belief that a sympathetic affinity existed between identical sounds and replicate things. It was believed that the name of a living being or an object was not just a simple or practical designation to facilitate the exchange of ideas between persons but that it was the very essence of what was defined, and that the actual pronouncing of a name was to create what was spoken. Similarly, the making of a recognizable copy of someone or something was to create an entity into which part of the spiritual personality of a living person or an object could be transferred. Both the act of speaking and the making of a material copy also provided the speaker or the artificer with a measure of control over persons and things. Purposeful utterance underlay all the rites which made use of formulae, just as the manufacture of images was designed to bring about reality through representation. It was this absolute faith in the creative power of sound and the productive force of representation which played so great a part in the daily lives of the Ancient Egyptians, that caused not only the men of the Classical World but their modern successors to conclude that Egypt was a land of the enchanters. The evidence for such a conclusion is not difficult to find, the meaning of the evidence has not always been understood.

As *Hike*, or Egyptian Magic, was essentially protective, it was used in the service of the State and the temples. In the service of the State it was employed to protect Egypt and the Pharaoh from the attacks of her enemies. Thus figurines inscribed with the names of nations and their rulers who were feared were made in order that they might be mutilated, trampled on, burnt and buried. In the temples the priests ceremonially cut to pieces the figure of Apepi, the fearful dragon-snake who daily endangered the Cosmic Order by attacking the boat of the sun at dawn and at sunset*. Everywhere spells were recited for the benefit of the sick and as a means of protecting the healthy from the attacks of the demons of disease. Likewise *Hike* was believed to be a protection against ghosts and accidents. It saved the dead from the demons who lurked in the Hereafter. It prevented the deceased from dying a second time from hunger when their living relatives neglected to provide their funerary offerings.

Not unnaturally the help of *Hike* was sought to gain some advantage for its user either in obtaining power over somebody or something, or as a love charm to gain some person's affection. In some instances its use was clearly an attempt to persuade the gods to bend the natural course of events to the user's benefit. Far less frequently attested are the instances where it was employed in what might be termed acts of black magic, performed secretly for personal ends. The most notorious of such abuses of *Hike* is recorded in the account of the attempt on the life of the Pharaoh Ramesses III (*c.* 1182–1151 BC), when a certain criminal obtained some of the royal magical texts and made wax figures and charms by which he hoped to cast a spell on the guardians of the King's harim. The repugnance felt by the Ancient Egyptians against the misuse of *Hike*

* See plate 14.

is illustrated by a story* contained in a Demotic papyrus which
relates how a certain Neneferkaptah tried to exploit for his own
ends the powers of *Hike* and how a series of misfortunes and
calamities overtook him and his family for his impious attempts.

What is sometimes called the Cult of the Dead with its seemingly disproportionate care lavished upon the deceased has
been largely responsible for the preservation of the greater
part of the physical remains of the Ancient Egyptian Civilisation. Were it not for the pictorial representations and the
inscriptions upon the walls of tombs, the texts upon sarcophagi
and coffins, and the contents of papyri containing various
recensions of *The Book of The Dead*, we should know very
little of what the Ancient Egyptians thought about the world
in which they lived. It would be a mistake to explain all the
care devoted to the furnishing of tombs, as one distinguished
archaeologist has done, by stating that the Ancient Egyptians
were in love with death. Nothing is further from the truth.
The fact is that they were in love with life, and with life as
they knew it in the Nile Valley. Their passion for the continuance of that life was such that they carried over into the
Hereafter many of the conceptions and ideas which they had
held about the physical world. But even so, their ideas about
the Hereafter were never very precise. Thus, in one form of
belief the dead man joined his ancestors who were already
lodged in the necropolis on the edge of the desert. There, if
his tomb was cared for by his living relatives, he could expect
to enjoy a happy and carefree existence, and even to revisit
some of the places which he had known when living. In other
accounts the dead man would pass to the fields of the blessed,
where he would continue to take everlasting pleasure in a way
of life not unlike that which he had known upon earth, but
immeasurably more serene and full of delight†. In yet another
account, what would seem to be a contrary conception held
that the soul of the dead man soared up into the sky to join
the solar bark and to be forever in the presence of the sun-god

* F. Ll Griffith, *Stories of the High Priests of Memphis.*
† See plate 13.

and the other deities who accompanied him. Difficulty is created in the modern mind by the fact that these differing ideas about the fate of the dead were all believed in at the same time. We have to remember that what appears to us to be contradictory the Ancient Egyptians accepted as complementary.

In what by necessity can be no more than an introduction to the Ancient Egyptian conception of cosmology the main facts have been presented to speak for themselves. No conscious attempt on the part of the author has been made to diminish or gloss over the considerable difficulties which the same facts invariably raise for the readers of a later age. Even a weighty treatise in the place of a single chapter could not succeed in minimising the task for either the author or his readers. If this introduction has opened a small window on to one aspect of a civilisation which was a 'very special blend of archaic tradition and the most advanced ideas' the author's purpose has been amply achieved.

Notes for Further Reading

This paper is printed in the form in which it was delivered as the introductory lecture in 1972. Because of the nature of the extensive evidence from Ancient Egypt no single work is able to present fully and systematically the religious beliefs of the Ancient Egyptians. For the purpose of further reading a reliable general introduction is J. Cerny, *Ancient Egyptian Religion* (London, 1952), containing a useful bibliography of the most important studies. A good, well-illustrated, introduction to the subject will be found in V. Ions *Egyptian Mythology* (London, 1968). As an introduction to the complexities of the religious thought of the Ancient Egyptians as expressed in their writings reference can be made to two excellent translations by R. O. Faulkner, *The Ancient Egyptian Pyramid Texts* (Oxford, 1969), and Aris and Phillips, *The Ancient Egyptian Coffin Texts*, Vol. 1 (Warminster, 1973). Translations of some of the relevant texts will also be found in J. B. Prichard, *Ancient Near Eastern Texts Relating to the Old Testament* (Princeton University Press, 1969).

2

The Cosmology of Sumer and Babylon

W. G. LAMBERT

Professor of Assyriology, University of Birmingham

Of the two centres of early civilisation, Egypt and Mesopotamia, Egypt in general excelled more in art, Mesopotamia more in technology and science. Further, the geographical setting of Egypt kept it more isolated from external contacts than Mesopotamia, and as a result Mesopotamia was more of an influence in the intellectual world of the times. Thus the geometry which Euclid systematised in his *Elements* was a Babylonian creation from the beginning of the second millennium BC. Knowledge of this had spread westwards, and the early chapters of Genesis similarly show Mesopotamian influence. Much later, in the period of the Persian Empire, astronomy was developed by Babylonian scholars, which the Greeks at first merely imitated and only finally surpassed in the first few centuries AD.

The area of these developments is more restricted than the modern term 'Mesopotamia'. It was confined to the southern end of the Tigris-Euphrates plain, roughly between modern Baghdad and Basra. It is a flat desert area, with marshes near the Persian Gulf, lacking wood and stone, which can sustain human occupation only by riverine irrigation. The period involved stretches from 3,000 BC to the time of Christ. Prior to 3,000 BC, this area was not remarkable in its Near Eastern context; indeed the mountains to the north were studded with

neolithic settlements. But under the culturally dynamic Sumerians this area sprang ahead just before the beginning of the
third millennium and produced a remarkable urban civilisation.
Writing was one of its first achievements, and Sumerian is now
largely understood, though it is related to no other known
language. Available materials were used, clay as the writing
material, and a reed stylus to impress the signs. The Sumerians
were in time swamped by Semites, who came down the Euphrates valley in successive waves, and the period of the second
and first millennia was Babylonian, culturally much indebted
to the Sumerians but in some particulars distinct. The final
death of this cultural tradition followed slowly on the Hellen-
isation of the East which resulted from Alexander's conquests,
but at least in Babylon itself a few families kept alive the
distinctive cuneiform script up to the end of the first century
AD.

The surviving sources for cosmology are both written and
pictorial, but mainly the former, and written material is usually
necessary to understand the pictorial. Most written remains
from Mesopotamia are administrative or economic in content,
and literary archives such as may contain cosmological material
are few. Figure 3 shows the chronological and geographical
distribution of these archives. The Babylonians merely followed the Sumerians, but in the second millennium there was a
spread of Babylonian culture to the north-west, and relevant
Babylonian texts have been recovered from outside the small
area of southern Mesopotamia. The Hurrians, who were spread
across northern Mesopotamia and Syria, had copies of Babylonian texts in the middle of the second millennium BC, and at
this same time there was also a Hittite archive which has yielded
the same kind of material from Anatolia. This can be grouped
with texts from Hurrian sites, since some at least of the Hittites'
cuneiform texts were passed on to them by the Hurrians. The
Assyrians on the upper Tigris, however, obtained their large
collections of Babylonian texts direct from the sources, and,
with the material of Hurrian and Hittite sites, they help to fill
the gaps between the surviving archives from Babylonian sites.

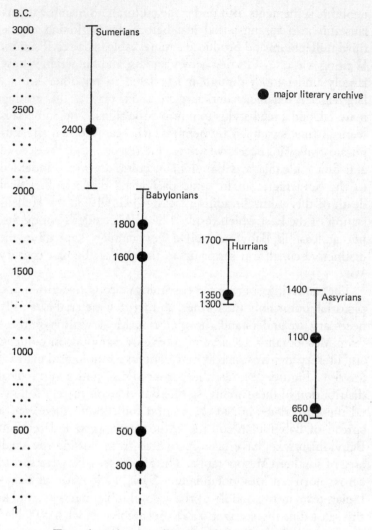

B.C.

3000 ··· Sumerians
··· ·
··· ·
··· ·
··· ·   ● major literary archive
2500
··· ·   2400 ●
··· ·
··· ·
··· ·
2000   ─┬─ Babylonians
··· ·
··· ·
··· ·   1800 ●
··· ·   1700 ─┬─ Hurrians
··· ·   1600 ●
1500
··· ·   1350 ●   1400 ─┬─
··· ·   1300 ─┴─   │ Assyrians
··· ·
··· ·   1100 ●
1000
··· ·
···· ·
··· ·
· ···   650 ●
500   500 ●   600 ─┴─
··· ·
··· ·   300 ●
··· ·   ┊
··· ·   ┊
1   ┊

3. Time chart showing the major finds of cuneiform texts

Despite the quantity of literary and scholarly texts from the various archives, and from finds of small groups or individual pieces, comparatively little is directly cosmological. A certain general idea can be gleaned from a multitude of texts of varying type: that the universe is formed of superimposed levels, the earth being roughly in the middle, the stars and heaven above, and cosmic water and the underworld below. However, formal statements of this concept are only known from the first millennium. Indeed, almost everything that would be considered cosmological from a modern standpoint is late and scanty. To obtain fuller insight into these ideas, and into the thought of the third and second millennia, myths of origin must be drawn upon. The *Just So* stories, which are perfect examples of etiological myths, can illustrate the use of this kind of material. If one wishes to know something about the giraffe's long neck, this story, which tells how the neck grew longer as the animal strained for the tender leaves at the tree tops, incidentally explains the function of the unusual part of this animal's anatomy. Similarly, myths about the origin of the universe not only explain how the universe was conceived to exist in the compiler's mind, but also in the course of telling how it came to receive this form they define the functions of its various parts.

It may appear that to quote a *Just So* story in the present context is meant humorously, but this is not so. If we smile at the modern story of how the giraffe got its long neck, we are ill-prepared to study ancient myth. Perhaps we are not altogether to blame, in that modern Western attitudes to myth are influenced largely by Classical Greek and Roman myths as known from works such as Ovid's *Metamorphoses*. This work comes from a sceptical age – Ovid himself wrote *expedit esse deos, et ut expedit esse putemus*, 'It is right that there should be gods, so let us believe there are' – and in any case his aim in writing this compilation was more literary than religious. This was inevitable in an age when myth had lost its original force, and no longer gripped the minds of men for its content. The same development happened in Mesopotamia too. By the first millennium the recording of narrative myths had virtually stopped, though older texts continued to be copied, and scholars drew on these and other myths in learned forms of theology. Even the surviving narrative myths from the third and second millennia in some cases show signs of the meaning being subordinated to the needs of the story-teller. No doubt the creative period of myth is to be set in prehistoric times, and confirmation can be found in the occurrence all over the Near East and Eastern Mediterranean of certain mythical motifs which are most naturally explained as products of diffusion rather than independent creation. For example, the idea that everything ultimately goes back to the prime element water occurs in sources Greek, Egyptian, Palestinian, Syrian, and Mesopotamian. The serious study of this material must not restrict itself to narrative myths, since however little our emotions may be moved by a myth that is told in one or a few sentences, these are no less important than the best products of the story-teller for the purpose of scholarly study.

While man in prehistoric times may have lacked scientific knowledge of the universe and its workings, he was as able as we are to observe and to think. With no developed material civilisation, he was much more conscious than we are of the natural forces around him. We can live with little care for the storm, unless it be unusually severe, and even the failure of a harvest need not cause us real fear of famine. Primitive man lived his whole life much more dependent on the natural forces around him; and lacking any appreciation of their interlocking, such as modern science has given us, he assumed that each force, be it the sun, moon, storm, power in the crops in spring and summer, or any other such part or aspect of the universe real or imagined, was an independent power. This much was the reasonable result of his observation. But depending on these powers as he plainly did, he went beyond this and assumed that these powers were in some sense like himself, and that they might be placated by his devotions. Also he asked questions about, for example, the origins of these powers, and answered in the form of stories which are now termed myth. An example can best illustrate the point. The changing seasons of growth and decay in nature were correctly observed and the spirit of growth in spring was personified as a deity. The explanation of the seasons was offered in a story which told how the goddess concerned went down to the underworld and came up again, so that plant life and animal procreation ceased during her absence. The Sumerian *Descent of Inanna* and the Babylonian *Descent of Ishtar* are very similar forms of this story, the Greco-Roman myth of Persephone or Proserpina is the same in the essential outline. The existence of two elements is very clear in this case: the correct observation of nature, and the myth which seeks to explain the phenomena.

This background to myth in general has been given because cosmology began as part of this process. The Sumerians had inherited the thought forms and numerous motifs from neolithic times, which they then developed and exploited in their own environment. This environment was physically distinctive as compared with the mountains in the North in its relative lack of rain and the unimportance of what fell for agriculture. This factor alone had its influence on cosmology. The idea that life arose from the marriage of heaven and earth is a widespread one, and often implies the shooting up of plant life as the result of father heaven sending down his rain into the bosom of mother earth. This idea occurs only rarely and marginally in Sumero-Babylonian texts. It is not the centre of any major text. It clearly stems from other areas which depended on rainfall for their plant life. In Sumer plants flourished on river water, spread by irrigation, and this water was known to have come ultimately out of the ground from springs. Since springs are not limited to any particular part of the earth, the Sumerians assumed that there existed a huge subterranean body of sweet water, called the Apsû, from which all springs draw their supplies. This reasonable idea was not an exclusive possession of the Sumerians, but they assigned greater importance than other peoples to this water and its patron god, Enki in Sumerian, Ea in Babylonian. The Egyptians both depended on riverine irrigation for their crops and believed in such a body of subterranean water. The relevant deity Nun, however, was not a major member of the Egyptian pantheon, and temples were not erected to him.

The earth of course overlaid the Apsû. This the Sumerians took to be a roughly flat disc, since it was self-evidently flat, and the movements of the heavenly bodies, and especially the sun, showed that it was limited in extent. The sun went down over the western horizon each night and rose over the eastern horizon the following morning. Clearly it went down under the earth during the night. Similarly the moon was conceived to go beneath the earth at certain times. This flat earth was the domain of Enlil, city god of Nippur, and the most powerful individual god of the Sumerian pantheon.

Heaven, synonymous with the sky, was conceived to be above the earth. However, while the passage of the heavenly bodies above the earth is clear for all to see, something more than observation is required for the concept of heavenly palaces in which the gods of heaven held court in anthropomorphic style. This idea probably derived from the personification of heaven itself rather than from the personification of the heavenly bodies. An, the Sumerian word for 'heaven', was the nominal head of the pantheon, and once he was conceived to reside up there in human style, though on a grander scale, the way was prepared to assuming that many other gods were similarly housed. The underworld likewise had its gods. This area of the universe, believed to be located beneath the earth, and to which human spirits went at death, was ruled by one god or goddess (according to which tradition is followed) associated with others. The correlation of the Apsû and the underworld, seeing that both were located in the lower parts of the universe, was potentially a problem. Sometimes the two ideas are blended, as when a watery character is assigned to the underworld, but often the underworld was conceived as a dry, dusty place; and if it was put on a different level it had to be below the Apsû, since that was immediately under the surface of the earth. This arrangement either contributed to, or arose from the idea that to reach the underworld it was necessary to cross a river, either the Ḫubur in Babylonian corresponding to the Greek Styx, or another body of water.

Thus the Sumerian view of the universe was one of cosmic levels. The evidence for it is not explicit, but derives from the use of terms which were inherited by the Babylonians, from whom there is more direct evidence. The picture is clear from many textual allusions firmly dated to the second half of the third millennium BC; and while there is much less evidence for the first half of the same millennium, and it is much less well understood, there can be little doubt that this conception of the universe goes back in Mesopotamia to at least 3,000 BC. It continued virtually unchanged until the end of Babylonian civilisation. In many ways it is a disappointment that the civilisation which produced so much in the collection of data and in the abstract sciences did not develop its cosmological ideas during the 3,000 years of its existence, and the reason for this can be sought. Two main factors were at work. One was the old, prehistoric mythological thinking, which, despite the fictional element in its products, was properly based on sound observation of the universe. This could have provided the basis for a scientific cosmology if conditions had been right. The other was the personification of natural forces into anthropomorphic gods and goddesses, and this led away from the realities of nature into theological fantasy. Unfortunately the growth of civilisation resulted in a diminution of the first of these two factors and an increase in the second.

The chief products of civilisation in Sumer and Babylonia were the cities. The population lived in them, and there was little village settlement. These cities were dominated by the temples. At first there was only one temple in a city, that of the chief deity of the place, but as the city increased in size, so did the temples, both in size and number. In any case the temple of the main deity of the town was the largest building, and in a flat landscape its prominence was all the more evident. Furthermore, the ancients did not distinguish between the world of nature and the products of the human race. It was held that the gods in council had decided to have cities as much as rivers: both alike were the result of divine decree. Thus there was no reason to leave cities out of cosmological thinking, and in the cities the gods obviously held the dominant position. The temples not only commanded theoretical allegiance, but they were also immensely powerful economic units. The presiding deities everywhere had their anthropomorphic elements stressed. They lived in 'houses' (there was no special word for 'temple'), where they were represented by statues in human form. They were fed twice a day, and clothed in certain rites. They had their spouses and children and a host of retainers, just like a human ruler, but on a bigger scale. In Sumerian times, at least, the city deity was conceived as the owner of the city, and the temple actually owned and worked much of the irrigated land, so that the ruler was much in the position of a farm bailiff managing the god's estates. With this kind of city, attention was directed away from the cosmic reality, which the deity represented, to the artificial world of anthropomorphic gods and goddesses.

Under this regime interest in cosmology flagged while much attention was paid to theological matters such as the ranking of the gods. The deities of all the cities were linked together into a kind of tribe, somewhat in the same way as were the Olympians in Greek mythology, consisting of three generations. The most senior gods were the most important, followed by the second generation and then by the third. Of the Babylonian myths certainly composed during the second millennium several are concerned with the relative ranking of the gods. The *Zû Myth* explains the rise of Ninurta within the pantheon. The modern title, *L'Exaltation d'Ishtar*, correctly explains the ancient myth so designated. The *Epic of Creation* explains how Marduk rose to supreme power among the gods, and there are fragments of a myth about the rise of Nabû, Marduk's son, to equality with his father. Cosmology in the proper sense occurs only incidentally and on the fringe of these myths.

A specific example will illustrate how cosmological ideas which are old and, for their period, reasonable lost their power in city theology. Many mythologies are concerned with ultimate origins, and in Mesopotamia it was customary to trace everything back to a single primeval element. It was believed that either Earth, Water or Time had alone existed in the beginning, and all else had somehow emerged from them.

These conceptions were as serious in their day as the speculations of Professor Hoyle and others are today, and it is almost
certain that they were part of a common Near-Eastern inheritance, since they turn up elsewhere so frequently. In the passage
of the centuries these prime elements were made to serve as the
ultimate ancestors of the most senior city gods.

Enlil, god of the city Nippur, exercised greatest power in the
Sumerian pantheon. That he had sons and daughters only
raised questions about his own ancestry. Whose son was he?
With other senior gods too genealogies were constructed to
answer this question. For Enlil the line begins with the pair
Enki and Ninki, and two examples out of many are given to
illustrate them:

(Old Babylonian)

Enki	Ninki
Enul	Ninul
Endashurimma	Nindashurimma
Endukuga	Nindukuga
Enlil	Ninlil

(Middle Babylonian)

Enki	Ninki
Enul	Ninul
Enmul	Ninmul
Enlu	Ninlu
Endu	Nindu
Enda	Ninda
Endim	Nindim
Enbuluh	Ninbuluh
Enpiriğ	Ninpiriğ
Engarash	Ningarash
Enshar	Ninshar
Enkur	Ninkur
Enamash	Ninamash
Enkingal	Ninkingal
Enkugal	Ninkugal
Enanna	Ninanna

Enutila	Ninutila
Endashurimma	Nindashurimma
Endukuga	Nindukuga
Enmesharra	Ninmesharra
Enlil	Ninlil

In all cases the first pair are the same, but the intervening pairs never agree completely between the various lists either in the number of pairs or in the distinctive element in each pair. These divergencies suggest that the intervening pairs were not in themselves important, but only serve to give remoteness to the first; and this conclusion is confirmed by the meanings of the names. In many cases the distinctive element in the pairs can be translated, but no cosmic sequence emerges when these meanings are taken together. However, with the first pair one should look more closely. *En* is Sumerian for 'lord' and *Nin* for 'lady'. *Ki* is Sumerian for 'earth'. The name in each case is an apposition, so they mean 'Lord Earth' and 'Lady Earth.' (This *Enki* is not the same as *Enki*, god of the Apsû, since the latter's name in its full form is *Enkig*.) The rationale of this genealogy is clear. Enlil had a spouse Ninlil. It was desired to make them descendants of the prime element Earth, so this latter was made into a marital pair Enki and Ninki, which was coupled by intermediate pairs to Enlil and Ninlil. The only mythology here is that Enlil descended from Earth.

An (Sumerian) or Anu (Babylonian) as the nominal head of the pantheon, like the president of a modern state, was similarly supplied with two ancestries. The one is a single line:

(Old Babylonian God List)

An	'Heaven'
Anshargal	'The-Whole-Heaven'
En-uru-ulla	'Lord-of-the-Primeval-City'
Urash	'Earth'
Bēlet-ilī	'Mistress-of-the-Gods'
Nammu	Nammu
Ama-tu-an-ki	'Mother-who-gave-birth-to-Heaven-and-Earth'

A brief exposition of this is needed. Since Anu heads it, it may be suspected that the order is reversed as compared with Enlil's, and this is confirmed by the appearance of 'Mother-who-gave-birth-to-Heaven-and-Earth' at the end, since this title suggests a primeval being. As interpreted by the present writer, Nammu heads the list chronologically and the title just quoted describes her. She is known elsewhere as the mother of Enki/Ea, and since he is the god of cosmic water beneath the earth, she too is presumably watery in character. The second generation is Urash, another Sumerian word for 'earth', and Bēlet-ilī, Babylonian for 'Mistress-of-the-Gods', seems to be an epithet of Urash. The third generation is En-uru-ulla, Sumerian for 'Lord-of-the-Primeval-City', which illustrates the cosmological importance of cities in Sumero-Babylonian thought. The remaining names, Anshargal and An, Sumerian for 'the whole heaven' and 'heaven', are either two generations, father and son, or two names of Anu. Evidence can be quoted for both of these possibilities. In this case all the stages of the genealogy are cosmologically important, and the final result is that Anu descended from cosmic water.

The other genealogy is bisexual, like Enlil's:

Dūri	Dāri	'Ever and Ever'
Laḫmu	Laḫamu	(Name of sea-monster)
Alala	Belili	(Uncertain)

(Powers invoked for exorcism)

This is no doubt to be read downwards, since in the Hittite *Kumarbi Myth* Alala is the predecessor of Anu. The first pair 'Ever and Ever' is a Babylonian phrase which is not grammatically masculine and feminine, but it is to be so construed for the purpose of this list. The second pair is equally artificial, since they are respectively the Babylonian and the Sumerian forms of a word indicating some kind of sea-monster. However, Alala and Belili are correctly male and female in mythology. As already stated, Alala's connection with Anu is known elsewhere, but while Belili is called 'the old lady' in Sumerian texts, she is not elsewhere so far known as wife of Alala or mother of Anu. Perhaps their mating was the work of the compiler of this list, who connected them because of the similarity of their name-type. The basic idea of this genealogy is that Anu descended from primeval Time.

The existence of two totally different ancestries of one god did not cause any crisis of belief in the ancient world. Both were accepted, and two examples of their combination into a single consolidated list are known. The one occurs in the Middle Babylonian god-list *An: Anum:*

Urash	Ninurash
Anshargal	Kishargal
Anshar	Kishar
Enshar	Ninshar
Duri	Dari
Laḫma	Laḫama
Ekur	Gara
Alala	Belili
Alala-alam	Belili-alam
En-uru-ulla	Nin-uru-ulla

Being part of a long compilation, this results from inclusion of everything relevant as was known to the author, and so far as it did not conflict with his theological ideas. Other names besides those of the two genealogies cited above have been used. The most important innovation is the placing of Urash and Ninurash at the head of the list: 'Earth' and 'Lady Earth'. This makes Anu the descendant of Earth, like Enlil, and it may well be the direct result of the ancestry of Enlil, since *An= Anum* contains both. The other combination of the two ancestries comes at the beginning of the Babylonian *Epic of Creation.* The names only are given here:

Apsû	Tiāmat
Laḫmu	Laḫamu
Anshar	Kishar

Anu
Nudimmud (Ea)
Marduk

Despite its brevity, the use of elements from both antecedents is clear. At the beginning Apsû and Tiâmat replace the watery Nammu with a pair. Apsû, the cosmic water under the earth, is masculine in Babylonian. Tiâmat is the ordinary Babylonian word for 'sea' and is grammatically feminine.* Once more then Anu descends from primeval water. The genealogy is extended by two generations to reach Marduk, whose rise is the subject of the poem.

This summary account of the theogonies of Enlil and Anu shows how the ancient concepts of prime elements were twisted and perverted to fit theological notions about the anthropomorphic city gods. So far as scientific cosmology is concerned they are worthless, and they help to explain why nothing really new arose in the course of Sumero-Babylonian history. However, the Babylonians did leave us a very few cosmological accounts which present their views more or less systematically. The *Epic of Creation* itself offers a combination of two traditional views in telling how Marduk, after defeating Tiâmat and so obtaining for himself supreme power in the universe, proceeded to organise it after his own taste. The following lines are from the end of Tablet IV:

135 Bel (i.e. Marduk) rested, surveying the corpse,
136 To divide the lump by a clever scheme.

137 He split her into two like a dried fish,
138 One half of her he set up and stretched out as the heavens.

139 He stretched a skin and appointed a watch,
140 With the instruction not to let her waters escape.

141 He crossed over the heavens, surveyed the celestial parts,
142 And adjusted them to match the Apsû, Nudimmud's (i.e. Ea's) abode.

143 Bel measured the shape of the Apsû,
144 And set up Esharra, a replica of the Eshgalla.

145 In Eshgalla, Esharra which he had built, and the heavens,
146 He settled in their shrines Anu, Enlil, and Ea.

Here Marduk makes the universe out of the body of the

---

* For the identification of *Tiâmat* and *Tehôm*, see Jacobs on p. 70.

defeated and dead Tiāmat. The upper part is turned into the heavens, and at this point Tiāmat, 'Sea', is definitely watery, since a skin was placed across the bottom of this part to prevent the water escaping. The completion of this scheme does not occur until line 62 of Table V: '[Half of her] he stretched out and made it firm as the earth.' This idea that earth and heaven were once joined and were separated as the first act of creation is alluded to briefly in three Sumerian texts, and once in the Hittite *Kumarbi Myth*. It is also found in widely scattered myths of other civilisations. Presumably then it was very old, and taken over in Mesopotamia from prehistoric tradition. However, the author of the *Epic of Creation* has not been too faithful to this tradition in that he has inserted a second version between his statements about the two halves of Tiāmat. Also the body thus sundered is not elsewhere a mass of water, so probably the author or a lost source on which he depended has joined the old myth on to the story of the defeat of Tiāmat. His second version is expressed in chiastic order in lines 145–6 of Tablet IV as quoted above. Three levels of the universe house the three gods who, before the rise of Marduk in the pantheon, formed a kind of trinity. In list form the gods and their corresponding levels are:

Anu – Heaven
Enlil – Esharra
Ea – Eshgalla

There has been much difficulty experienced in the interpretation of this idea, but a careful study of the whole section and other occurrences of the terms in the *Epic of Creation* clears up the problems. Marduk started on his work of reorganisation of the universe with two things: the Apsû and Tiāmat's body. The Apsû, the primeval male water, was killed by Ea right at the beginning of the story, in Tablet I 65–72. After the killing, Ea set up his abode on the dead body. This is an etiological account, explaining how Ea came to be patron god of the cosmic waters beneath the earth, as he traditionally was. So Marduk here uses this pre-existing Apsû as a pattern on which were made the other two levels in the three-decker version. As explained in lines 141–2, heaven was trimmed to match the Apsû. The term Eshgalla, Sumerian for 'big house', is used elsewhere for the underworld, but since the *Epic of Creation* nowhere even alludes to such a place its author was free to use this term for the lower cosmic water. In this text it is then a synonym of the Apsû, as is clear from the parallelism of lines 143–4 and from its relationship to Ea in the following couplet.

Esharra is more difficult. Since nothing was to hand to serve, with modifications, as Esharra, Marduk had to 'build' it. It is Enlil's level (the name is used elsewhere for his temple in Nippur), and it was somewhere between heaven and the Apsû. One immediately thinks of the earth, and this might seem to be confirmed by the cosmology assumed in the *Atra-ḫasis Epic*. This is also based on three levels: Anu in heaven, Enlil on the earth, and Ea in the Apsû. However, other passages in the *Epic of Creation* which mention Esharra do not permit of its being earth. In Tablet V lines 119–22 Marduk tells his fathers of his plan to build Babylon at the central point of his universe:

> Above the Apsû, the emerald (?) abode,
> Opposite Esharra, which I built for you,
> Beneath the celestial parts, whose floor I made firm,
> I will build a house for my luxurious abode.

The central position of Esharra between heaven and Apsû is taught here too, but Babylon is to be built 'opposite' and not 'upon' it. So it cannot be the earth here. Furthermore, when Babylon had been built and Marduk's temple, Esagil, had been completed for him as a present from the other gods, Marduk sat in state in his new home:

> He sat in splendour before them,
> Surveying its 'horns' at the base of Esharra.
>
> Tablet VI 65–6

The 'horns' of Esharra are its pinnacles, and if these were more or less level with the base of Esharra, that can only be heaven in some sense. This might seem an impossible conclusion, but the deduction is inevitable from the text of the passage, and the concept of a temple's highest points jostling with heaven is a commonplace in Sumero-Babylonian literature. The apparent difficulty is solved in that the Babylonians had a doctrine of several superimposed heavens, and in the *Epic of Creation* Esharra is the lower heaven. The text to be considered next will confirm this. Thus the *Epic of Creation* combines two originally separate cosmologies: the one is of two levels: heaven and earth, obtained in this story by the splitting of Tiāmat's body. The other, taken over from a tradition also attested in *Atra-ḫasīs*, was of three levels – heaven, earth and Apsû – occupied by Anu, Enlil and Ea. This the author perverted somewhat by putting Enlil in a lower heaven when he combined it with the double-decker version. He wanted the earth for Marduk, his hero.

The fullest exposition of the levels of the Babylonian universe is contained in a non-literary text. Two late compilations of miscellaneous paragraphs contain it, but one offers only a short form:

> The upper [heavens] are of *luludānītu*-stone, of Anu.
> The middle [heavens] are of *saggilmut*-stone, of the Igigi.
> The lower heavens are of jasper, of the stars.
> *(AfO* 19 p. 110 iv 20–2)

The fuller form adds further interpretation of the heavens, and corresponding earths:

> The upper heavens are of *luludānītu*-stone, of Anu.
>   He settled the 300 Igigi therein.
> The middle heavens are of *saggilmut*-stone, of the Igigi.
>   Bel sat therein on the lofty dais in the chamber of lapis
>   lazuli, he lit a lamp of *elmēšu*-stone.
> The lower heavens are of jasper, of the stars.
>   He drew the constellations of the gods thereon.
> [On the] base of the upper earth he made frail mankind lie down.
> [On the] base of the middle earth he settled his father Ea
>   [. . .] . he did not distinguish . . .
> [On the base] of the lower [earth] he shut in the 600
>   Anunnaki [. . .] . . . [. . .] . jasper.
> *(KAR* 307 obv. 30–8)

In this text the Igigi are the gods of heaven, the Anunna

the gods of underworld. The short version is clearly original, since the extra lines of the long version plainly contradict it. The denizens of the three heavens according to the two sets of lines are:

upper:	Anu	The 300 Igigi
middle:	The Igigi	Bel (i.e. Marduk)
lower:	The stars	The constellations

The extra lines are probably extracts from another text, since the first one lacks the subject, Marduk, who is only mentioned for the first time in the second extra line. The short version is also the older, since Marduk's exaltation only took place late in the second millennium BC, and this is presumed in the extra lines, but not in the original description of heaven. The original version with Anu having a special heaven of his own is attested in other cuneiform texts. For example, during the flood, according to Tablet XI of the *Gilgamesh Epic*, the gods in terror fled up to the heaven of Anu, i.e. to the highest spot in the universe to get as far away from the disaster on earth as possible. The later version was in fact known to the author of the *Epic of Creation*, but he uses different terms. The distinctive thing about this one is that the chief god resides in the middle heaven. In the *Epic of Creation* the terminology is different in that the upper and middle levels are called simply 'heaven' and 'Esharra', and the stars, though duly mentioned, receive no designation to indicate the level in which they move. The placing of Enlil in Esharra is an earlier form of the tradition which put Marduk in the middle heaven, since Enlil was the chief god before Marduk usurped his position.

The three 'earths' require no comment. 'Earth' in Babylonian can mean both 'earth' in the English sense and 'underworld'. Thus the three levels are the abode of men, the Apsû, and the underworld. Six levels, then, are the fullest form of the Babylonian universe, three heavens and three 'earths', as they would appear in a side view of the universe. A picture of the earth as conceived by the Babylonians from a top view survives on a Late Babylonian tablet in the British Museum, and there is accompanying text (Plate 15). The 'Mappa Mundi' shows a circular earth surrounded by water called the 'Bitter River'. Beyond this water lay a number of triangular areas called 'regions', probably eight in number when the text was complete. These remote areas were for special men, for example the hero of the flood story, who had been made immortal. The two lines running down the centre of the round earth are probably meant as the Tigris and Euphrates, and the horizontal band into which they run is marked as 'marsh', i.e. the marshes at the head of the Persian Gulf. The oblong above the centre has the name 'Babylon' inscribed in its right-hand end, but it is not clear if this is really in place. Other cities marked are in circles or ovals, and the city Babylon did not in fact cover both Tigris and Euphrates. Only the latter flowed through it. One would like to think that the compass hole in the middle is Babylon, but then the purpose of the oblong is unexplained. There are in fact many obscure points about this map and its accompanying text. Most of the surviving text concerns the 'regions' beyond the 'Bitter River', but one section lists mythical sea monsters. These were presumably considered to swim in the 'Bitter River' or perhaps under the earth as well. Although as preserved the text does not make this clear, the author must surely have accepted the idea that the water around the earth also extended beneath it.

Study of the stars, contrary to a common misunderstanding, was not a developed science in Mesopotamia until the Persian and Hellenistic periods. Previously their astronomy was primitive. But suddenly it flourished and developed on purely scientific lines and even the Greeks were at first happy to learn from it. So far as it went it was a remarkable achievement in the history of mankind, but in one point it was deficient. Without any understanding of a round earth, and without the need to grapple with the resulting problems, since all the work for this astronomy was conducted in one small area of southern Mesopotamia, a system was constructed which could only work within this area. The Greeks, being spread in Hellenistic times over a much wider area, and being familiar with the notion of a spherical earth, extended the Babylonian system to take care of these extra complications, and in so doing completely surpassed it.

In a sense, then, the Babylonian cosmological ideas were a dead end. An overwhelming interest in city theology had deprived them of any desire to pursue cosmology in a scientific manner, and the ideas of the stone age were passed on virtually unchanged. However, the human race both collectively and individually most commonly learns by first making mistakes. The Babylonians were a tremendous stimulus in the world of their time, and the failure of their ideas was the background against which others reached more lasting conclusions. The Greeks, by handling the notion of primeval elements impersonally, were able to move on to abstract, scientifically biassed cosmology. And the Hebrews with their monotheism provided a theistic view of the universe which has better stood the test of time.

In conclusion, it may be well to indicate the lack of sound evidence for two ideas commonly attributed to the Babylonians and their forerunners. First, the allegation that the Babylonian universe was conceived as a kind of ziggurat going up to a peak, and secondly the idea that the sky was thought to be in the shape of a dome. The only surviving Babylonian world map presents a flat world in general with one very small area marked 'mountain' on the northern edge of the circular disc adjacent to the surrounding water (*see* Plate 15). The Epic of Creation, as quoted on p. 55, teaches that the Apsû, the earth and heaven were all made by Marduk to match, so indicating levels of the universe of equal size and the same shape. There are, it is true, some allusions to the concept of a cosmic mountain, but these occur in literary and poetic contexts and it is not possible to reconstruct a precise image from them. The most explicit ones speak of a mountain in the East from which the sun-god rises every morning, and since the phenomenon was seen on the horizon the term 'mountain' cannot be taken too literally.

The idea of a vault of heaven is not based on any piece of

evidence. P. Jensen, whose *Die Kosmologie der Babylonier* of 1890 was for its time an excellent work, simply translated the Babylonian word for 'heaven' in Enūma Eliš IV 145 by 'vault of heaven' (p. 288–9) and thereafter assumes that the point is proved. Further support has also been sought in the word *pulukku*, which Oppert compared with the Arabic *falakun* 'celestial sphere' (*see* W. Muss-Arnolt, *A Concise Dictionary of the Assyrian Language* p. 807). However, the idea is no longer held. W. von Soden's *Akkadisches Handwörterbuch* (p. 879) gives the meanings of this word as: needle, boundary post, boundary. Thus to the Babylonians the universe consisted of superimposed layers of the same size and shape separated by space. This is confirmed in that a cosmic cable was thought to hold the various levels together, to prevent them drifting apart, and movement between the different levels was (at least for gods) achieved by use of a cosmic staircase.

Notes for Further Reading

There is no major work on Sumero-Babylonian cosmology that is even generally trustworthy in the present state of knowledge. Most aspects will be treated in detail in the present writer's forthcoming *Babylonian Creation Myths*, but for the present the following works may be consulted. J. B. Pritchard (ed.), *Ancient Near Eastern Texts Relating to the Old Testament* (Princeton University Press, latest edition 1969) offers translations of the following texts quoted or referred to: *The Descent of Inanna* (52 ff.), the *Epic of Creation* (60 ff.), the Babylonian *Gilgamesh Epic* (72 ff.), the *Descent of Ishtar* (106 ff.), the *Zŭ Epic* (111 ff.) the *Kumarbi Myth* (120 ff.). *L'Exaltation d'Ishtar* is available in German translation of B. Hruška, *Archiv Orientální* 37 (1969) 347 ff. The ancestries of the senior gods are discussed by J. J. A. van Dijk in *Acta Orientalia* 28 (1964) 1 ff. The rise and spread of the exact sciences in the ancient world is described by O. Neugebauer in *Proceedings of the American Philosophical Society* 107 (1963) 528 ff.

Plates

1. The universe as conceived by the ancient Egyptians (from Piggott (ed.) *The Dawn of Civilisation* (London, 1961).
2. Map of the constellations. This tomb ceiling painting reveals that in common with other peoples of antiquity the Ancient Egyptians divided the night sky into groups of stars and assigned to each group a symbolic form and name.
3. Nun holding up the Solar Barque. Representations of Nun, the primeval being, are comparatively rare. In this picture from the papyrus of Hunefer in the British Museum he is shewn holding aloft the Solar Barque in which the sun is represented by the scarabaeus beetle, Khopri. Because the word *khopri* was very similar in sound to the word meaning 'to come into being, to exist' the Ancient Egyptians here used the principle of the *rebus* (British Museum).
4. The Solar Barque. The disc of the sun is accompanied by the figure of the sun god Re, in the form of a mummified man wearing the head of a hawk surmounted by a disc and holding the *'ankh*, the symbol of life. In this form, under the name of ReHerakhte, the deity was worshipped as the Rising Sun from Ions, *Egyptian Mythology* (London, 1968).
5. The Primeval Goose. Above the Primeval Goose stands the Ram, the symbol of Amun, the principal deity of Thebes, the capital city of Egypt during the New Kingdom (British Museum).
6. Nut the sky goddess. From the inner face of the lid of the sarcophagus of Queen Ankhnes-Neferibre in the British Museum. According to one account, Nut became the mother of the sun god, Re. It was supposed that daily at evening she swallowed the sun's disc and gave birth to it the next morning. From Ions, *Egyptian Mythology* (London, 1968).
7. Ptah, represented as a mummified man, was the god of the city of Memphis At Memphis he was considered to be the creator of the world. An ancient tradition credited him with the invention of crafts. Artisans claimed his as a protecting deity and his high priest bore the title 'Lord of the Master-Craftsmen', from Ions, *Egyptian Mythology* (London, 1968).
8. Thoth. Thoth is shown as a man wearing the head of an ibis. Associated with the moon, Thoth represented various systems of reckoning. He was also credited with the invention of writing and as the keeper of records he became the law-giver of the gods (Greenfield Papyrus, British Museum).
9. Shu separating Nut from Geb. In this representation from the Greenfield Papyrus in the British Museum, Shu, the god of the atmosphere, raises the body of his daughter Nut, the sky goddess, above the recumbent body of his son Geb, the earth-god. On each side of Shu stand figures of Khnum, a creator god of life and living things, who was also the guardian of the sources of the Nile at the First Cataract (British Museum).
10. The eight genii of Hermopolis. The genii are represented in four pairs, the masculine with the heads of frogs and the feminine with the heads of serpents. Beginning from the right they are named above in hieroglyphs as *ḥḥ* and *ḥḥt* (endlessness), *kk* and *kkt* (darkness), *n'w* and *n't* (invisibility), and *ỉmn* and *ỉmnt* (invisibility).
11. The young sun emerging from the sacred lotus (Papyrus of Ani, British Museum).
12. Maat, the personification of the conception of the essential equilibrium and harmony of the universe, is represented as a woman, usually sitting and

wearing on her head an ostrich feather. The feather alone was frequently used to write her name. She also came to represent the ideas of truth and justice. In the texts she is called the daughter of the sun god, Re, from Ions, *Egyptian Mythology* (London, 1968).

13. The fields of the Blessed. In this picture from the Greenfield Papyrus the deceased lady Nesitanebtashru is seen in the fields of the Blessed, punting a boat, ploughing with a yoke of oxen, reaping, adoring the Benu bird, the symbol of abundance, and sitting before heaps of barley and wheat. The serpent headed boat in the lower register is the Boat of Millions of Years in which Re sailed over the sky daily, accompanied by those souls who had been declared justified in the Great Judgement (British Museum).

14. The slaying of Apepi. The 'Great Cat', a very ancient solar deity, 'who is in Heliopolis' is depicted cutting to pieces the evil demon serpent, Apepi, the enemy of the sun god, at the foot of the sacred tree (From the Papyrus of Hunefer in the British Museum).

15. The Babylonian Mappa Mundi (British Museum/B.M. 92687).

16. Relief from stone tablet recording benefactions to the cult of the sun god Shamash in Sippar, a town in southern Mesopotamia, from the reign of Nabù-apla-iddina, king of Babylon *c.* 850 BC. The scene generally is explained by the caption in the blank upper left-hand portion: 'the statue of Shamash, the great lord, who resides in the temple Ebabbara, which is in Sippar'. While the purpose is partly theological, a cosmological setting is given, although not altogether consistently. Shamash sits on an elaborate stool on the right while on the left two human figures approach, probably the king led by a priest, and behind them there is a goddess. The top right-hand caption explains some cosmology: 'Sin (the moon god), Shamash and Ishtar (= Venus) are placed opposite the Apsû between the Destiny-decreeing gods'. The symbols of these three deities occur in the same order just below the caption, and the Apsû is represented by wavy lines at the bottom of the picture. The 'Destiny-decreeing gods' seem to be the two figures above the pillar who hold the large solar disc on a stand by one rope each. Probably the the pillar is a symbolic part of the actual temple structure of Ebabbara, but the thing rising behind Shamash and then bending over above his head is hardly meant as any architectural feature, but seems to be a support for the gods who are holding up the solar disc.

To understand these particulars it is necessary to know that in literary texts temples are often spoken of as resting on lower parts of the cosmos and vying with heaven in their pinnacles. Modern readers tend to dismiss this as poetic imagery only, but the ancients seem to have taken it literally. Since temples were the biggest buildings their foundations did go deep and their pinnacles high. Here then we see a temple, represented by one pillar only, resting on the Apsû and its capital level with moon, sun and Venus. Then, out of scale, the artist has presented the solar disc as suspended between two gods who were supposed to escort it in this way across the sky. The caption suggests that all three heavenly bodies were thought to cross the sky similarly, but only one is so presented, and that in addition to its appearing in the row with the other two. The caption level with the head of Shamash reads: 'Turban of Shamash, *mushshu* of Shamash', but the meaning of *mushshu* is unknown (from L. W. King, *Babylonian Boundary Stones and Memorial Tablets in the British Museum* (London, 1962).

17. Impression of Mesopotamian cylinder seal from Tell Asmar, *c.* 2300 BC. The main scene consists of a god-boat (a boat the prow of which merges into a god) being punted over water while the sun god (identifiable by the rays

3

5

4

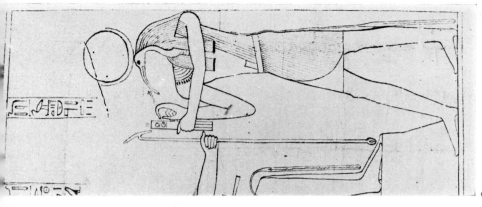

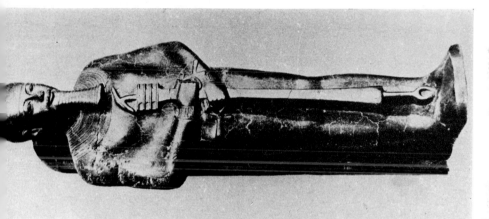

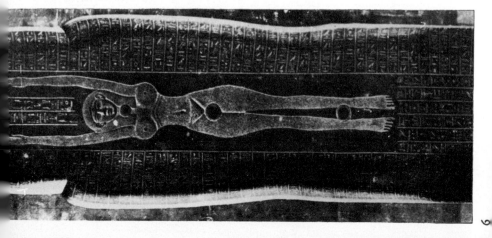

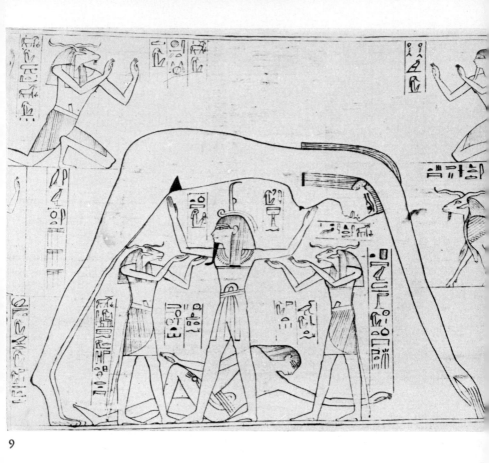

9

10

12

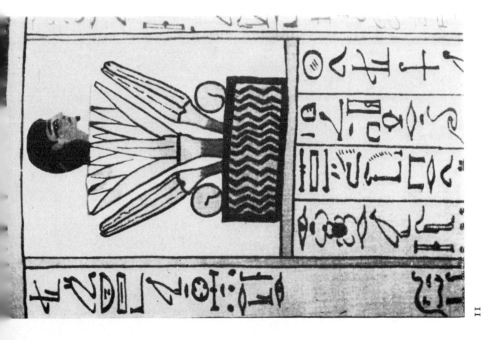

11

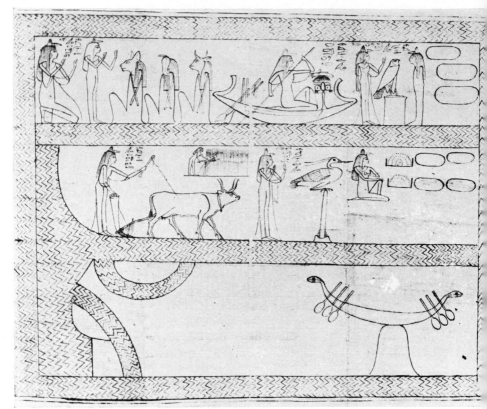

13

14

15

16

17

rising from his shoulders) is seated steering. The idea followed rationally from the sun's going down over the western horizon each evening and rising in the east each morning. Obviously he passed under the earth during the night, and since it was believed that cosmic water existed there, a boat was the obvious means of travel (from H. Frankfort, *Stratified Cylinder Seals from the Diyala Region* (Chicago, 1955).

3
Jewish Cosmology*

RABBI LOUIS JACOBS
Lecturer in Talmud, Leo Baeck College, London

A study of the Jewish sources demonstrates that the Jews did not develop in any period of their history a special cosmology of their own. They adopted or accepted the cosmologies of the various civilisations in which they lived, but utilised these for the religious purposes with which they were primarily concerned. Judging by the classical Jewish writings, Jewish preoccupation was with the God of cosmos not with the cosmos itself. There was, to be sure, a profound interest in natural phenomena but chiefly as pointers to God who initiated them and whose glory was revealed through them.

> Lift up your eyes on high,
> And see: who hath created these?
> He that bringeth out their host by number,
> He calleth them all by name;
> By the greatness of His might, and
>   for that He is strong in power,
> Not one faileth'   (Isa. 40:26)

> The heavens declare the glory of God,
> And the firmament showeth forth His handiwork
>   (Ps. 19:2)

The vivid description of the universe and its creatures in Psalm 104 begins with:

> O Lord my God, Thou art very great;
> Thou art clothed with glory and majesty

---

* A brief explanation of some of the technical expressions used in this chapter will be found on pp. 85–6.

It is preferable, therefore, to speak not so much of Jewish cosmology as of cosmologies that have been entertained by Jews. For these it is necessary to examine the picture of the world as portrayed in the Bible and the Rabbinic literature, in medieval Jewish philosophy and the *Kabbalah*, with a glance at later Jewish thought. The consideration of medieval notions in a series of chapters concerning ancient cosmologies is, I think, justified when we observe that many of these notions are themselves ancient. Even when they are stressed particularly by the medieval thinkers, they go back in the main to the period with which this series is concerned.

Although the Biblical writings extend over a period of several hundred years, the cosmological picture in these writings is remarkably uniform. One can, without distortion, refer, therefore, to the 'Biblical' view and quote in its support passages from the different books of the Bible. A moot point first to be noted is whether the Biblical record knows of the concept of a cosmos. T. H. Gaster[1] suggests that to the ancient Hebrews 'the world was not an organic unity but a collection of disparate phenomena individually controlled and collectively disposed at the will and pleasure of their common Creator'. There is, in fact, no word in the Bible for 'universe' or 'cosmos'. The word *olam*, later (in the Rabbinic literature, for instance) meaning 'world' or 'universe', means, in the Bible, 'eternity', with the possible exception of the use of this word in the late book of Ecclesiastes (3:11). However, the use of 'very good' at the end of the creation narrative in Genesis (1:31), as opposed to the simple 'good' in which the details of the creation are described, does suggest that, over and above the excellence of each particular, the writers had a concept of the excellence of the cosmic order as a whole.[2] In any event, by the Rabbinic period (i.e. from the beginning of the civil era to c. 500) the idea of a cosmos is well established. God is there frequently spoken of as 'King of the Universe' (*melekh ha-olam*).

The Biblical picture is clearly geocentric. The earth has the shape of a flat disc[3] so that if one were able to travel far enough

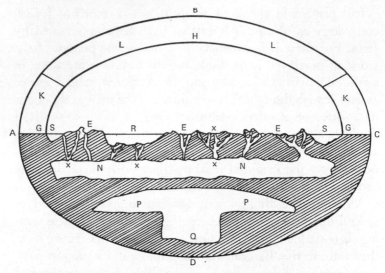

Explanatory Key:
ABC = the upper heaven; ADC = the curve of the abyss; AEC = the plane of the earth and seas; SRS = various parts of the sea; EEE = various parts of the earth; GHG = the profile of the firmament or lower heaven; KK = the storehouses of the winds; LL = the storehouses of the upper waters, of snow, and of hail; M = the space occupied by the air, within which the clouds move; NN = the waters of the great abyss; xxx = the fountains of the great abyss; PP = Sheol or limbo; Q = the lower part of the same, the inferno properly so called.

4. Heaven, the earth and the abysses
(from W. F. Warren, *The Earliest Cosmologies* (New York, 1909))

one would eventually arrive at the 'ends of the earth' (Deut. 13:8; 28:64; Isa. 5:26; Ps. 135:7). This term can simply refer to far-distant places, but its use is evidence of the cosmological picture. The 'corners' or 'wings' (*kanefot*) of the earth (Isa. 11:12; Ezek. 7:2; Job 37:3) may be a synonym for the 'ends of the earth'. If, on the other hand, the earth is not conceived of as a disc but as a square strip, the 'corners' may be understood literally. It is also possible that the term *kanefot* refers to the four directions, north, south, east and west. The earth rests on pillars (Job 9:6). Stretched above the earth is the sky, 'heaven' (*shamayim*) or 'firmament' (*rakia*), a solid substance[4] (Gen. 1:6–8) resting on pillars (Job 26:11).[5] Just as the earth has an 'end' so does the sky (Deut. 4:32). The sun, moon and stars are positioned in, or just beneath,[6] the firmament (Gen. 1:

Biblical conception of the world: 1. waters above the firmament; 2. storehouses of snows; 3. storehouses for hail; 4. chambers of winds; 5. firmament; 6. sluice; 7. pillars of the sky; 8. pillars of the earth; 9. fountain of the deep; 10. navel of the earth; 11. waters under the earth; 12. rivers of the nether world.

5. Biblical conception of the world

(from N. M. Sarna, *Understanding Genesis* (New York, 1966))

14–17) and they move across it (Ps. 19:1–7). Beneath the earth is *Sheol*,[7] the abode of the dead (Num. 16:28–34; 1 Sam. 28:13–15; Isa. 14:9–11; Eccles. 9:10). There are waters above the firmament (Gen. 1:6–7) as well as beneath it. Some of the waters beneath the firmament were gathered together at the beginning of creation to form the seas (Gen. 1:9–10) but, in addition, these waters flow beneath the earth (Exod. 20:4; Deut. 4:18; Ps. 24:2) where they are connected to the waters of *Tehom*, the great deep (Gen. 1:2). Fountains, wells and springs flow from these waters beneath the earth. The Deluge was caused by a tremendous outpouring of the fountains of *Tehom* as well as by the opening of the windows of heaven (Gen. 7:11). Rain is produced by the clouds (Gen. 9:11–17; Job 26:8; Eccles. 11:3). The water in the clouds comes from the waters above the firmament so that when the heaven is 'shut up' there is no rain (Deut. 11:17) while when the 'good treasure' of heaven is opened the rain falls in abundance (Deut. 28:12).

While the word 'heaven' (*shamayim*) is used of the firmament or sky, it is also used of the area located above the waters that are above the firmament. This area is also known as 'the heaven of heavens' (Deut. 10:4; I Kings 8:27). This is normally the abode of God (Exod. 20:19; Isa. 66:1; Ezek. 1:1).

It is clear that this cosmological picture owes much to the general ancient Mesopotamian cosmologies, especially the Babylonian. Here we note that, while in some Biblical passages the mythological elements derived from the ancient cosmologies are still very prominent, in the creation narrative in Genesis[8] there appears to be a conscious effort to suppress them. Possibly, traces of such mythological themes as the battle between the gods and the dragon of chaos in the deep are present even in the Genesis narrative in the use of the word *Tehom* (= the Babylonian *Tiāmat**) without the definite article for the great deep (Gen. 1:12) and in the use of the plural 'Let us make man' (Gen. 1:26) reflecting the counsel of the gods. But these are largely matters of vocabulary or usage only (*cf.* the modern use of 'Wednesday' and 'Thursday' for the days of the week). The narrative as a whole breathes the spirit of monotheism. In other Biblical passages (Isa. 27:1; 30:7; 51:9–10; Hab. 3:8; Ps. 74:13–14; 89:10–11; 93; Job 3:8; 9:13) the references to the ancient myths are far more pronounced. The 'myth and ritual' school has even purported to detect an annual re-enact-

* For this term see p. 55.

ment of the primordial conflict on the New Year festival in ancient Israel.[9]

It has also been suggested that the Temple was constructed on the parallel of the world, e.g. the Holy of Holies corresponds to the heavens, the outer house to the earth, the laver to the sea and so forth.[10] God has provided man with a home and man in gratitude provides God with a place in which He can reside and which mirrors the home of man.

Nowhere in the Biblical record is the doctrine of *creatio ex nihilo* clearly mentioned. Although the root *bara*, 'to create', is used only of God's activity, never of man's, it does not in itself imply *creatio ex nihilo*; indeed, the root meaning seems to be that of 'cutting out' i.e. of an existing material. As Abraham Ibn Ezra (1089–1164)[11] pointed out in the Middle Ages, the root *bara* is used in the Genesis narrative not alone for the original creation (Gen. 1:1) but also for the later creation of the sea-monsters (Gen. 1:21) and of man (Gen. 1:27). The earliest reference in Jewish literature to *creatio ex nihilo* is in the second book of Maccabees (late second to early first century BC; *see* 7:28). In the Wisdom of Solomon (first century BC to first century AD), on the other hand, creation is out of 'formless matter' (11:17).

On the whole the cosmological picture as it appears in the vast Rabbinic literature is not very different from the Biblical picture. The world of nature was thoroughly familiar to the Rabbis and they introduced special benedictions to be recited when a man observes its marvels. There are benedictions on observing the sea, mountains, comets, thunder and lightning, strange creatures and trees in bloom.[12] But side by side with this there is a strong attempt to discourage speculation on cosmic origins and on those cosmic matters that are beyond human experience, in all probability because of the heretical, especially dualistic, views, which could follow from these.[13] Ben Sira (early second century BC) is quoted in the Talmud[14] for his, according to the Rabbis, sound advice: 'Do not pry into things too hard for you or examine what is beyond your reach. Meditate on the commandments you have been given;

what the Lord keeps secret is no concern of yours.'[15] (Eccles.
3:21–22). Thus the *Mishnah*[16] states: 'Whosoever reflects on
four things, it were better for him if he had not come into the
world – what is above; what is beneath; what is before; and
what is after.' In the comment of the Jerusalem Talmud to
this passage in the *Mishnah* it is said that the *Mishnah* follows
the opinion of Rabbi Akiba (*c.* 50–150 BC) but that according
to Rabbi Ishmael (early second century AD) it is permitted to
'expound the work of creation'. In any event there are to be
found in the Rabbinic literature discussions on the manner of
God's creation and the nature not alone of the terrestrial but
also of the celestial realms.

The School of Shammai (first century AD) held that heaven
was created first and the earth afterwards. The School of Hillel
(first century AD) held that the earth was created first and
afterwards the heavens. But the Sages held that heaven and
earth were created simultaneously.[17] We are told[18] that a
philosopher said to Rabban Gamaliel (first century AD): 'Your
God is a great craftsman, but He found good materials to help
Him in the work of creation, namely, *Tohu* and *Bohu*, darkness,
wind, water and the deep', to which Rabban Gamaliel retorts
that these, too, were created by God and he quotes scriptural
verses in support. This is the Rabbinic equivalent of the discussion concerning *creatio ex nihilo*. As late as the third century,
however, the Palestinian teacher Rabbi Johanan could say that
God took two coils, one of fire and the other of snow, wove
them into each other and created the world.[19] According to one
Rabbinic theory all things were created simultaneously on the
first day of creation but made their appearance at different
stages in the other six days, just as figs are gathered simultaneously in one basket but each selected in its time.[20] The
opinion that the primordial light was a garment with which
God wrapped Himself before creation[21] is probably a reference
to a theory of emanation which became especially prominent
in the *Kabbalah*.[22] The idea is found that God created several
worlds and destroyed them before creating this one.[23] God is,
as it were, proud of the world He has created. He declares that

His creation is 'very good' (Gen. 1:31). If the Creator praises His wonderful works who would dare to criticise them?[24] There was a belief in Rabbinic times that originally the sun and the moon were the same size but that because the moon protested that 'two kings cannot wear the same crown' God told her to make herself smaller.[25]

The fondness of the Rabbis for descriptions of the immense size of the universe has undoubtedly an apologetic motivation. The aim is either to praise God or to defend Israel's worth in creation. It is, in fact, difficult to know how far these statements were intended to be taken literally. For instance, a sage declares, in opposition to his colleagues, who say that the world rests on twelve or on seven pillars, that the earth rests on one pillar and its name is 'Righteous', for it is said: 'But *Righteous* is the foundation of the world' (Prov. 10:25).[26]

Similarly, a heavenly voice is made to taunt Nebuchadnezzar when he said: 'I will ascend above the heights of the clouds; I will be like the Most High' (Isa. 14:14). The heavenly voice replies: 'Man has only seventy years in which to live.' But the distance from the earth to the firmament is a journey of five hundred years, and the thickness of the firmament is a journey of five hundred years, and likewise the distance between one firmament and the other. Above them (the seven firmaments) are the holy living creatures:

> the feet of the holy living creatures are equal to all of them together; the ankles of the living creatures are equal to all of them together, the legs of the living creatures are equal to all of them; the knees of the living creatures are equal to all of them; the thighs of the living creatures are equal to all of them; the bodies of the living creatures are equal to all of them; the necks of the living creatures are equal to all of them; the heads of the living creatures are equal to all of them; the horns of the living creatures are equal to all of them. Above them is the throne of glory: the feet of the throne of glory are equal to all of them; the throne of glory is equal to all of them. The King, the Living and Eternal God, High and Exalted, dwelleth among them. Yet thou didst say: 'I will ascend above the heights of the clouds, I will be like the Most High'![27]

(Ezek. 1:5)

Again, when Israel is apprehensive that God has forgotten her, He replies (significantly in terms taken from the Roman army):

My daughter, twelve constellations have I created in the firmament, and for each constellation I have created thirty hosts, and for each host I have created thirty legions, and for each legion I have created thirty cohorts, and for each cohort I have created thirty maniples, and for each maniple I have created thirty camps, and to each camp I have attached three hundred and sixty-five thousands of myriads of stars, corresponding to the days of the solar year, and all of them I have created only for thy sake, and thou sayest, Thou hast forgotten me and forsaken me![28]

As for the time the universe will endure, Rabbi Kattina (third century AD) said that the world will endure for six thousand years and it will be desolate for a thousand, but Abaye (early fourth century) said that it will be desolate for two thousand years.[29]

The Rabbis believed in the possibility of miracles happening, seeing in miracles not a suspension of natural or universal law (of which there was no such conception in their thinking) but, as they put it, a 'change in the order of creation'. They believed that miracles did not only occur in the past but occur also in their own day, although there are differences of opinion whether it was praiseworthy or otherwise for a miracle to be performed on behalf of a contemporary. Revealing in this connection is the bizarre anecdote about a man whose wife died, leaving him with a babe for whom he was unable to afford a nurse. A miracle was performed for him and his breasts became as a woman's that he might suckle his child. One of the Rabbis said: 'How great this man must have been that such a miracle was performed for him!' But another Rabbi said: 'On the contrary! How unworthy this man must have been that the order of creation was changed on his behalf!'[30]

In a well-known Rabbinic passage it is said that ten things were created on the eve of the first sabbath of creation in the twilight, among them: the mouth of the earth (Num. 16:32);

the mouth of the well (Num. 21:16); the mouth of the ass
(Num. 22:28); the rainbow, the manna, and the rod (Exod.
4:17).[31] Similarly, it is said that when God created the sea He
imposed a condition on it that it be divided before Israel, as
He did with the fire that it should not harm the three young
men, with the lions that they should not harm Daniel, and
with the fish that it should vomit out Jonah.[32] With the exception of the remark about 'universal law', which, as we have
noted, is anachronistic when applied to the thought of the
Rabbis, Zangwill's[33] explanation comes close to the meaning
of these passages: 'The Fathers of the *Mishnah*, who taught
that Balaam's ass was created on the eve of the Sabbath, in the
twilight, were not fantastic fools, but subtle philosophers,
discovering the reign of universal law through the exceptions,
the miracles that had to be created specially and were still a
part of the order of the world, bound to appear in due time
much as apparently erratic comets are.'

It is extraordinary how the ancient creation myths reappear
in the Rabbinic, especially the Midrashic, literature. Thus,
while it is stated that the Leviathan was created on the fifth
day, together with the other fishes,[34] the fins of the Leviathan
are said to radiate such brilliant light as to obscure the light of
the sun.[35] The Leviathan is said to be the plaything of God.[36]
There are references to a male and female Leviathan, God slaying the female.[37] But, interestingly enough, the conflict with
the Leviathan is projected into the future. At the end of days
the angels will engage the Leviathan in combat without success
and eventually it will be slain by Behemot, and its flesh will be
fed to the righteous.[38] The mythological *motif* is similarly
pronounced in the legends which tell of the rebellion of the
Prince of the Sea at the time of creation.[39] The astonishing
feature in all this is that the mythological passages are late,
dating from the Amoraic period (third century onwards) not
from the earlier Tannaitic period. It would seem that the
Mesopotamian creation myths lived on among the people and
were at first refused any recognition by the official Rabbinic
teachers.[40] Mythological *motifs* of very ancient vintage similarly

re-emerge in the Kabbalistic literature from the thirteenth century.

Medieval Jewish cosmology, generally speaking, is the standard Greek cosmology in its Arabic garb. The central problem for the Jewish thinkers in this area was the doctrine of *creatio ex nihilo*. All the Jewish thinkers reject the Aristotelian view that matter is eternal but while Maimonides (1134–1205) and the majority of these thinkers have an unqualified belief in *creatio ex nihilo*, considering this to be a cornerstone of the Jewish faith, Gersonides (1288–1344) adopts the Platonic view

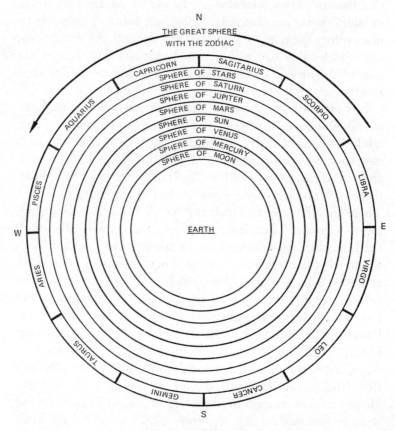

6. Maimonides' cosmology.

of a formless matter, existing from all eternity, upon which God imposed form.[41]

Maimonides devotes the opening sections of his great Code to a description of the universe in his conviction that man's contemplation of the vastness and the marvels of God's creation would evoke his sense of awe and lead eventually to the love and fear of God.[42] For Maimonides[43] there are three types of being in the universe: (1) beings having both form and matter but who suffer decay, such as humans, animals, plants and minerals; (2) beings having form and matter but which do not suffer decay, such as the spheres and the heavenly bodies attached to them; (3) beings that are non-corporeal, having only pure form, such as the angels. There are in all nine spheres. The nearest of these is the sphere to which the moon is attached. In ascending order there are then the spheres of Mercury, Venus, the Sun, Mars, Jupiter and Saturn. Above these is the eighth sphere to which all the other stars are attached and above all these is the great ninth sphere which revolves each day from east to west and through its revolutions the other spheres revolve. The spheres are translucent so that when seen from the earth all the stars appear to be attached to a single sphere. But each of the eight lower spheres is subdivided into many other spheres 'like the layers of an onion' (the comparison is made by Maimonides), some of these revolving from east to west, others from west to east. The spheres are both colourless and weightless. The blue appearance of the sky is an optical illusion.[44] The ninth sphere is divided into twelve sections each named after the planet situated beneath it. These are the twelve signs of the Zodiac.[45] Some of the stars seen in the sky are smaller in size than the earth, some of them larger. The earth is forty times larger than the moon but the sun is 170 times larger than the earth. The smallest of the stars is Mercury and none of the stars is larger than the sun.[46] The stars and spheres are intelligent beings who offer praises to their Creator.[47] All sublunar beings are composed of the four elements, fire, air, water and earth.[48] Maimonides concludes:[49]

'When man reflects on these topics and comes to recognise all creatures, from the angels and the spheres to human beings like himself, and when he observes the wisdom of the Holy One, blessed be He, as manifested in all things and in all creatures, his love for God grows, his soul thirsts and his flesh longs to love God, blessed be He. Such a man is filled with awe and dread at the thought of his own lowliness, poverty and insignificance when compared with one of the great and holy bodies to say nothing of one of the pure, disembodied spirits, so that he becomes aware of himself as a vessel full of shame and confusion, empty and lacking.'

Although the medieval Jewish thinkers believed in miracles there is a marked tendency to interpret these as uncommon but natural phenomena. For Gersonides, for example, the regularity of nature is itself the most powerful evidence of God's work. Miracles only occur when there is a special and pressing need to demonstrate God's power. All miracles are the result of the Active Intellect, the mediator between the higher Intelligences, which move the heavenly spheres, and the human intellect. The Active Intellect only operates, therefore, in the sublunar world. Furthermore, miracles are only a temporary, never a permanent, interruption of the natural order, which latter is guaranteed by the orderly movements of the heavenly bodies not subject to the influence of the Active Intellect. It follows that no miracle can ever occur in the realm of the spheres and hence the Biblical passages which seem to say that the sun stopped for Joshua and the shadow moved back for Hezekiah have to be understood otherwise than appears on the surface.[50]

The idea, which goes back to the Greeks, of a close correspondence between man, the microcosm, and the universe, the macrocosm, was utilised by some of the medieval Jewish thinkers but was virtually ignored by others.[51]

In the *Kabbalah* the theory of emanation is the central feature. *Creatio ex nihilo* means, for the *Kabbalah*, the emergence of 'somethingness' out of God's *Nothingness*. Remarkably reminiscent of Far Eastern cosmogonic theories, is the Zoharic comparison of the way in which the *Sefirot*, the creative powers or potencies in the Godhead, emerge from *Ein Sof*, the Limitless, the unknown and unknowable Ground of Being, to the silkworm which spins its cocoon out of itself.[52] In Hasidic thought, strongly influenced by the *Kabbalah*, the simile is changed to that of the snail 'whose garment is from itself' and applied to the world which is God's garment.[53] In some versions of Hasidism this results in a completely acosmic view. From God's point of view, as it were, there is no cosmos at all. The cosmos only enjoys existence from the point of view of God's creatures.[54]

The main concern of the *Kabbalah* is, in any event, not with the physical universe but with the 'upper worlds'. Thus the Rabbinic saying regarding God creating worlds and destroying them is referred, in the Lurianic *Kabbalah*, to the creative processes in the Godhead in which the 'vessels' of the *Sefirot* were at first shattered because they were too weak to contain the splendour of the light of the limitless (*Ein Sof*). In fact, for the *Kabbalah*, the details of the cosmic order as perceived by man are no more than a pale reflection in the physical world of the spiritual entities and their various combinations on high.[55]

The ancient theory of cosmic cycles (*Shemmitot*) won much support in the early *Kabbalah* but was eventually repudiated. The theory, as it appears in the *Kabbalah*, runs that there are time cycles each lasting six thousand years followed by a thousand year sabbath. There are seven of these cycles in all culminating in the great Jubilee after 49,000 years have passed. In one version the whole process begins afresh after the Jubilee. Again in some versions the daring view was put forward that each cycle has its own *Torah*. Thus we are now living in the cycle governed by the *Sefirah* 'Judgement' and the *Torah* we have is one that is adjusted to such a situation. Therefore our *Torah* contains negative as well as positive precepts. But in the cycle of 'Lovingkindness' a different *Torah* prevails containing only positive precepts. It was this idea, in flat contradiction to the dogma of the immutability of the *Torah*, that caused the later Kabbalists to reject the whole doctrine.[56] But the doctrine was resurrected by more recent post-Darwinian thinkers in a somewhat forlorn attempt at coping with the problems raised for believers by the evolutionary theories and the new picture of the great age of the earth.[57]

Modern Jewish thinkers, with few exceptions, adopt the view that the nature of the physical universe is to be investigated by the methods of science and that it is not a matter of religious faith; so that for these thinkers there is no Jewish cosmology any more than there is a Jewish mathematics. Even a completely traditionalist thinker like Rabbi A. I. Kook accepts, for example, the theory of evolution in his contention that the creation narrative in Genesis belongs to the 'secrets of the *Torah*' and hence must not be taken literally. With a strong resemblance to the views of Teilhard de Chardin, Kook believes that an evolutionary theory is in the fullest accord with the basic optimism of the *Kabbalah* of which he was an adherent.[58]

The new picture of the universe revealed by modern science has produced hardly any new theological speculations among Jews but some little consideration has been given to the problems raised by space travel and the possibility that there are intelligent and moral beings on planets other than earth. In the encyclopedia of human knowledge compiled by Rabbi Phineas Elijah Hurwitz of Vilna (d. 1821), entitled *Sefer Ha-Berit*,[59] there is speculation on this theme as early as the beginning of the nineteenth century. Hurwitz[60] believes, on the basis of Isa. 45:18, that there are creatures on planets other than earth. He refers to the Talmudic passage[61] in which, according to one opinion, *Meroz* (Judges 5:23) is a star and yet, says Hurwitz, *Meroz* is cursed for not coming to the help of the Israelites, which indicates that it is inhabited. Hurwitz goes on to admit that the creatures on other planets may have intelligence but refuses to believe that they are endowed with freewill, for this, he argues, is only possible for creatures with a human constitution.

More recently Rabbi Gunther Plaut[62] asks: 'Will the possibility that there are intelligent creatures on other planets impose any strain on our religious beliefs?' He replies: 'The modern Jew will answer this question with a firm "No". An earlier generation, rooted in beliefs in an earth-centered universe, might have had some theological difficulties, but we have them no longer. That God should, in His vast creation, have caused only one earth and one manlike genus to evolve, is in fact harder to believe than that His creative power expressed itself in other unfathomable ways. This does not in any way diminish our relationship to Him or His to ours. Just as a father may love many children with equal love, so surely may our Father on high spread His pinions over the vastness of creation.'[63] A more detailed and acute examination of the problem is that given by Rabbi Norman Lamm under the title: 'The Religious Implications of Extraterrestrial Life'.[64] Among other matters, Lamm discusses whether Judaism holds the doctrine of man's cosmic significance to be a cardinal principle of the Jewish faith.

To sum up, the Jews never invented a cosmology of their own. Still less has there been any official Jewish cosmology dictated by Jewish orthodox belief. But certain cosmological themes, deriving from Babylonian, Greek or Arabic sources, have been stressed or rejected according to the doctrinal bent of individual Jewish thinkers in the various periods of their history.

Notes

1. 'Cosmogony' in *International Dictionary of the Bible*, Vol. I, 702–9.
2. See U. Cassuto, *From Adam to Noah*, trans. I. Abrahams, Jer. (1961), 59.
3. This is possibly the meaning of the *ḥug* ('circle') of the earth in Isa. 40:22. Kimhi and many moderns, however, understand the *ḥug* of the earth to be the vault of the sky. But it is worth noting that in Job 22:14 there is a reference to the *ḥug* of heaven, which would suggest that the verse in Isaiah refers to the earth.
4. In Exod. 39:3 the root *raka* is used for beating gold into thin plates.

F

When the 'heaven' fails to produce rain it is described as being like iron (Lev. 26:19). No significance is to be attached to the dual form *shamayim* (as in *yadayim*, 'hands', *raglayim*, 'feet') since this word, like the word for 'water', *mayim*, is really a plural and not a true dual form. The correct translation is 'heavens' but *shamayim* is used in our verse as a synonym for *rakia* which is the singular form. Cf. Gesenius, *Grammar*, 88:1.

5. The earth, too, rests on pillars, Job 9:6. Presumably these pillars were at either end of the sky and earth. In the diagram in *Interpreter's Bible*, op. cit., 703 (also reproduced in Nahum M. Sarna's, *Understanding Genesis*, McGraw-Hill, New York, 1966, 5), the pillars of earth and heaven are depicted as two huge masses reaching to and supporting the sky at either end with the earth crossing them in the middle, but I have been unable to discover the evidence to warrant this.

6. The expression used is 'God *set* (or *put*) them' (*va-yitten*) in the 'firmament' (Gen. 1:17). See S. R. Driver, *The Book of Genesis* (London, 1926), 9–11.

7. In the diagram referred to, the 'waters beneath the earth' are depicted as dividing the earth and *Sheol*, *Sheol* being beneath these waters. But from Num. 16:28–34 it appears that *Sheol* is immediately beneath the earth's surface with nothing else in between.

8. There are two different creation narratives. The first (Gen. 1:1–2:4a) is ascribed by the critics to 'P' (sixth century BC), the second (Gen. 2:4b–25) to 'J' (tenth century BC).

9. The 'myth and ritual' school was inaugurated by S. Mowinckel's *Psalmenstudien*. Cf. Snaith's *The Jewish New Year Festival* (1947), and the bibliography in O. Eissfeldt, *The Old Testament An Introduction*, trans. Peter A. Ackroyd (Oxford, 1966), 110, note 29.

10. See R. Patai, *Man and Temple* (London, 1967). The belief that Palestine is at the exact centre of the earth is probably referred to in the 'navel' of the earth (*tabbur ha-aretz*) in Ezek. 38:12. (But in Judges 9:37, where the same term is used, the reference is simply to the centre of that district.) Cf. the Rabbinic legend that when David began to dig for the site of the Temple the waters of *Tehom* welled up and threatened to engulf the earth (*Makkot* 11a).

11. *Commentary* to Gen. 1:1.

12. See *Mishnah Berakhot*, Ch. 9, and the Babylonian Talmud on this.

13. See the remarkable illustration in the Jerusalem Talmud (*Ḥagigah* 1:2) of the king whose palace was erected over the sewers. The honour of the king demands that no one is allowed to inquire as to what was there before the palace had been erected. Cf. H. Albeck's *Commentary to the Mishnah*, Jer.-Tel-Aviv, (1964), *Moed*, Supplementary Note to *Ḥagigah* 2:1, 510–11.

14. Babylonian Talmud, *Hagigah* 13a. Cf. *Genesis Rabbah* 8:2, Theodor-Albeck (ed.), 58 and Theodor's note 1.

15. New English Bible Version.

16. *Ḥagigah* 2:1. The terms *le-fanim* and *le-aḥor* used in this passage of the *Mishnah* can either mean (as appears from the Babylonian Talmud and is generally assumed) 'before' and 'after' in time or they can mean 'before' and 'after' in space, i.e. what is beyond the confines of the earth. *Cf. Genesis Rabbah* 1:10.

17. Babylonian Talmud, *Ḥagigah* 12a.

18. *Genesis Rabbah* 1:9.

19. *Genesis Rabbah* 10:3.

20. *Genesis Rabbah* 12:4.

21. *Genesis Rabbah* 3:4.

22. See A. Altmann, *Studies in Religious Philosophy and Mysticism* (London, 1969), 128–39.

23. *Genesis Rabbah* 9:2.

24. *Genesis Rabbah* 12:1.

25. Babylonian Talmud, *Ḥullin* 60b.

26. Babylonian Talmud, *Ḥagigah* 12b.

27. Babylonian Talmud, *Ḥagigah* 13a. The number seven is the general Semitic sacred number. While this passage is not necessarily the earliest reference to the 'seven heavens' in the Rabbinic literature (for further references see Ginzberg, Legends of the Jews, Philadelphia, 1942, Vol. V, pp. 10–11, note 22), the concept appears only from the third century. This lengthy note of Ginzberg should be consulted on the whole question as well as his notes 20 and 21 on p. 19. In note 20 Ginzberg remarks that the use of the apparent plural form for *shamayim* (see note 4 above) led to the conception that the idea of several heavens is already met with in the Bible.

28. Babylonian Talmud, *Berakhot* 32b.

29. Babylonian Talmud, *Rosh ha-Shanah* 31a; *Sanhedrin* 97a.

30. Babylonian Talmud, *Shabbat* 53b.

31. *Avot* 5:6. The other four are: the Shamir-worm, used by Solomon miraculously to cut the stones for the Temple; the letters of the Two Tablets of Stone on which the Decalogue was engraved; the engraving tool for these; and the Two Tablets themselves.

32. Genesis *Rabbah* 5:6.

33. Zangwill's comment is quoted by J. M. Hertz, *Commentary to the Daily Prayer Book* (London, 1947), 687f.

34. Babylonian Talmud, *Baba Batra* 74b.

35. *Pesikta de Rab Kahana*, Buber (ed.), p. 188a.

36. Babylonian Talmud, *Avodah Zarah* 3b; *Baba Batra* 74b.

37. Babylonian Talmud, *Baba Batra* 74b.

38. *Pesikta de Rab Kahana*, Buber (ed.), pp. 188 a–b.

39. Babylonian Talmud, *Baba Batra* 74b. On the whole question of these ancient myths in the Rabbinic literature see the lengthy note by

Ginzberg, *Legends of the Jews* (Philadelphia, 1942), Vol. V, note 127, pp. 43–6.

40. On this see E. Urbach, *Ḥazal*, Jer. (1969), pp. 169f.

41. See Saadia, *Emunot Ve-Deot*, I:1–15; Maimonides, *Guide for the Perplexed*, II, 13–25; Albo, *Ikkarim*, I, 23; Gersonides, *Milḥamot*, VI.

42. *Yesodei Ha–Torah* 2:2.

43. *Yesodei Ha–Torah* 2:3.

44. *Yesodei Ha–Torah* 3:1–3.

45. *Yesodei Ha–Torah* 3:6.

46. *Yesodei Ha–Torah* 3:8.

47. *Yesodei Ha–Torah* 3:9.

48. *Yesodei Ha–Torah* 3:10–11.

49. *Yesodei Ha–Torah* 4:12.

50. See Gersonides *Commentary* to Joshua 10:12.

51. See I. Broydé in *Jewish Encyclopedia*, Vol. VIII, 544–5.

52. Zohar I, 15a.

53. Jacob Joseph of Pulnoye, *Toledot* (ed. Warsaw, 1881), 39a; *Keter Shem Tov* (ed. Jer., 1968), 12a. The actual wording about the snail is found in *Genesis Rabbah* 21:5 but in an entirely different context.

54. This view is especially pronounced in the Habad school of Hasidism and more particularly in the writings of R. Aaron Hurwitz of Starosselje, see my study of R. Aaron's thought, *Seeker of Unity* (London, 1966).

55. See G. Scholem, *Major Trends in Jewish Mysticism*, 3rd ed. (London, 1955), 205–86.

56. An excellent account of the doctrine of cosmic cycles in the *Kabbalah* is I. Weinstock's *Studies in Jewish Philosophy and Mysticism* (Heb.) (Jer., 1969) 151–241.

57. See Weinstock, *op. cit.*, 230–41.

58. See Kook's *Orot Ha–Kodesh* (Jer., 1938), Part IV, 19–22.

59. 2nd ed. (Warsaw, 1881).

60. Part I, *Maamar* 3, Chs 2–4, 30–2.

61. Babylonian Talmud, *Moed Katan* 16a.

62. *Judaism and the Scientific Spirit* (New York, 1962), 36–9.

63. In a note (p. 79) Plaut observes: 'There is some reason to believe that even the Jewish ancients were already hinting at a wider view. Judaism knows various expressions for God. It calls Him "King of the World" and also, "King of All Worlds". A *Midrash* states that before our earthly creation God created and destroyed many worlds (Gen. R. 3:7).' But, of course, the ancients had no notion of 'worlds' inhabited by non-human, intelligent beings (other than angels), and the reference in the *Midrash* is to God creating and destroying many worlds *before* the creation of this one.

64. In *Tradition*, Vol. 7, No. 4–Vol. 8, No. 1, Winter 1965–Spring 1966, 5–56.

Explanation of Technical Terms

1. AMORAIM. Amoraic; see TALMUD.

2. LURIANIC KABBALAH. The mystical doctrine evolved by the great Jewish mystical teacher and poet Isaac Luria (1534–72), associated especially with the Galilean town of Safed, where Luria settled not long before his death.

3. MIDRASH. Interpretation of Scripture. The term is also used for the considerable non-legal part of the anonymous Rabbinic literature. Some of this is dated to the Tannaitic (Mishnaic) period, but most of it is later, some as late as the twelfth and thirteenth centuries. The Midrashic literature contains a widely ranging assortment of Rabbinic teachings, many of which do not strictly fall under the heading of exegesis.

4. MIDRASH RABBAH ('Great Midrash'). A collection of *Midrashim* on the Pentateuch and the Five Scrolls (Ruth, Lamentations, Esther, Ecclesiastes and the Song of Songs). In fact these are quite separate compilations produced at widely differing dates. Although based on the Biblical books, they mostly consist of homilies, legends and edifying religious teachings.

5. MISHNAH ('teaching'). A codification of mainly legal teachings of the early Rabbis (Tannaim). In its present form the Mishnah is generally assumed to be the work of the Patriarch Rabbi Judah (end of the second century AD), but it contains a good deal of much earlier material and some slightly later authorities are mentioned in it. For a translation, see H. Danby, *The Mishnah* (Oxford 1933).

6. PESIKTA DE RAB KAHANA. A collection of midrashic homilies for certain special sabbaths and the festivals, named after the teacher who is mentioned at the beginning of the work. It is probably seventh-century, but contains much early material.

7. TALMUD ('teaching'). Compendium of the teachings of the later Rabbis (Amoraim), presented in the form of a running commentary on the Mishnah. There are two Talmuds, the Babylonian and the Palestinian (or Jerusalem) Talmud, which are quite distinct although they contain much material in common. The Palestinian Talmud was compiled in the Rabbinic schools of Palestine in the later fourth century; the Babylonian Talmud (often referred to simply as 'The Talmud', in consequence of the wider currency it has enjoyed) was compiled in the Babylonian schools in the late fifth century, but was somewhat edited and added to in the following century.

8. TANNAIM. Tannaitic; see MISHNAH.

9. TORAH. The Pentateuch or 'Five Books of Moses'. Sometimes applied to Scripture as a whole, to the Biblical law, or to religious teaching in general.

10. ZOHAR, THE ('The Book of Splendour'). The classic text of the medieval Jewish mysticism (Kabbalah). The principal part of the Zohar, written in a peculiar form of Aramaic, sets out the mystical doctrine under the guise of discussions on parts of the Bible by Rabbi Simeon bar Yoḥai and other second-century Palestinian Rabbis. In fact it was composed in Spain towards the end of the thirteenth century by Moses de Leon.

11. *Translations of Rabbinic texts*:

The Mishnah, tr. H. Danby (Oxford, 1933).

The (Babylonian) Talmud (London (Soncino) 1935–52).

Midrash Rabba (London (Soncino), 1939).

The Zohar (London (Soncino), 1934).

Jerusalem Talmud, French tr. by M. Schwab (Paris, 1932–33).

4

The Cosmology of Early China

JOSEPH NEEDHAM

Master of Gonville and Caius College, University of Cambridge

This chapter[1] is to be on the cosmology of ancient and medieval China; and I think we may fairly give about a third of it to the astronomical aspects and two-thirds to eschatology. How did scientific minds interpret the visible universe, and what did scholars as well as ordinary people think could happen to them after death?

The period of the late Warring States and the Earlier and later Han (from the − fourth to the + second century) was a time of intense speculation in cosmology and great advances in observational astronomy. By about + 180, when Tshai Yung described them, three schools of thought had become current; these were the *kai thien* school with its doctrine of the domed universe or 'heavenly cover'; the *hun thien* school, which drew out the celestial circles of the astronomical sphere much as was done in other civilisations; and finally the *hsüan yeh* or 'infinite emptiness' school, which had its ups and downs but was very prevalent throughout Chinese history, and which one might describe as the recognition of infinite empty space. The *hun thien* people, of course, did not commit themselves as to what the heavens were made of, or what it was that the stars were attached to, but the *hsüan yeh* scholars were quite clear that it was a matter of infinite empty space, so that they were really very far advanced for their time.

On internal evidence the *kai thien* theory has to be regarded as the most archaic of the three. The heavens were imagined as a hemispherical cover and the earth like a bowl turned upside down, the distance between them being 80,000 *li*.[2] The precise distances were based on rather unsophisticated calculations, but we are presented with two concentric domes, as shown in the diagram. Rain falling upon the earth among the

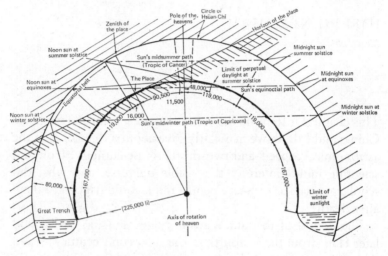

7. Reconstruction of the *kai thien* cosmology

dwellings of men, the *oikoumene* (οἰκουμένη), flowed down to the four edges to form the rim-ocean in the great trench. The heavens were thought to be round, but the earth square, and the vault of heaven rotated like a mill from right to left carrying with it the sun and moon, which nevertheless had a proper motion of their own from left to right but much slower than that of the vast wheel to which they were attached. The rising and setting, however, of the heavenly bodies was considered to be an illusion because they really never passed below the base of the earth, so the sun was essentially regarded as a circumpolar star, illuminating continually different parts of the earth's surface as if by a kind of searchlight beam. But its distance from the pole varied according to the season, following one of the roads that passed between seven parallel declination circles. Chatley,[3] who studied the *kai thien* universe theory very carefully, remarked that there was just about enough physical truth in the scheme to render it acceptable to rather archaic geometers having not much more than the Pythagoras theorem at their disposal. But I believe, under correction from the Assyriologists, that a somewhat similar double-vault theory of the world existed in Babylonia, so it might have been one of those culture-traits that passed both west and east, westwards to the Greeks and eastwards to the Chinese, to be developed later in both civilisations into the theory of the celestial sphere.[4] However, it was characteristically Chinese to say that the heavens were circular and the earth square. That was part of the Yin and Yang ideology, and arose naturally from the four cardinal points of earthly space and the circles (or apparent circles) of the celestial sphere.

The *hun thien* school, that of the celestial sphere itself, was based on the universal experience of an apparent spherical motion around the earth's centre. This idea had developed among the Pre-Socratics in Greece and came to be particularly associated with Eudoxus of Cnidus, who died in − 356. The earliest exponent of it in China whose name has come down to us was Lo-hsia Hung, who died in − 104, and the oldest description comes from the pen of Chang Hêng (*ca* + 100) in his book called *Ling Hsien* of which we have fragments still. The title means 'The Spiritual Constitution of the Universe', and in this text he described the basic celestial spheres of all astronomy in a very clear text. In another book, his 'Commentary on the Armillary Sphere', the *Hun I Chu*, he is rather precise; he says that 'the heavens are like a hen's egg, and as round as a crossbow bullet. The earth is like the yolk of the egg and lies alone in the centre. Inside the lower part of the heavens there is water. The heavens are supported by vapour, and the earth floats on the waters.'[5] Then he goes on about the circumference of the heavens, and the number of its degrees. But Chang Hêng's words are interesting because he attributes the visualisation of the celestial sphere to a time long before his own, and shows how the conception of a spherical earth with antipodes would naturally arise out of it. The *hun thien* school contributed in an important way to the growth of scientific instrumentation. By the time of Lo-hsia Hung armillary rings for celestial measurements had fully developed, and their combination into the armillary sphere for measuring the positions of stars and planets in the heavens was probably complete in the century before Chang Hêng.

Lastly, the third school, that of the *hsüan yeh* doctrine, was associated with another name, a man about whom we know very little, Chhi Mêng, who flourished during the later Han, that is to say between o and about + 200. He might have been a younger contemporary of Chang Hêng but we do not know his exact dates. Later on a remarkable scholar, Ko Hung (ca. + 300), the greatest alchemical writer in all Chinese history but also interested in astronomy, wrote an interesting passage which is preserved in the *Chin Shu* (History of the Chin Dynasty). He said that

'the original books of the *hsüan yeh* school were all lost, but Chhi Mêng, one of the imperial librarians, remembered what its masters before his time had taught concerning it. They said that the heavens were entirely empty and void of substance. When we look up at them we can see they are immensely high and far away, without any bounds. It is like seeing yellow mountains sideways at a great distance, for then they all appear blue; or when we gaze down into a valley a thousand fathoms deep, it seems sombre and black. But the blue of the mountains is not a true colour, nor is the dark colour of the valley really its own. The sun, the moon and the company of the stars float freely in empty space, moving or standing still, and all of them are nothing but condensed vapour. The seven luminaries sometimes appear and sometimes disappear, sometimes move forward and sometimes retrograde, seeming to follow each a different series of regularities. Their advances and recessions are not the same. It is because they are not rooted to any basis that their movements can vary so much;

they are not in any way tied together. Among the heavenly bodies the Pole Star alone always keeps its place, and the Great Bear never sinks below the horizon in the west as do the other stars. The speed of the luminaries depends on their individual natures, which shows they are not attached to anything, for if they were fastened to the body of heaven, this could not be so.'[6]

It is obvious, then, that the *hsüan yeh* conception was a very enlightened one, a vision of infinite space with celestial bodies at rare intervals floating in it. It was really more enlightened than the Aristotelian-Ptolemaic conception of concentric crystalline spheres, which was dominant in European thought for a thousand years or more. The *hsüan yeh* school pervaded Chinese thought to a greater extent than some have believed, because we find it often cropping up in later centuries. It might well be argued that the lack of deductive geometry in China – there was nothing corresponding to Euclidean geometry there – was helpful in this case, because the lack of planetary theory was probably a result of this lack of deductive geometry. One might even argue that the Greeks had too much, since the apparent mathematical beauty of cycle on epicycle, orb on orb, eventually came to constitute rather a strait-jacket, posing unnecessary difficulties for Tycho Brahe, Giordano Bruno, Copernicus and Galileo.

One also ought to add that Buddhism, which began to take root in China from the + second century, contributed a certain amount to the *hsüan yeh* theory of the Chinese as time went on. It is well known that the cosmological views of Buddhist philosophy were always enlightened and spacious. There is a famous text where monks asked of the Buddha how far distant Jambū-dvipa was from the Brahma world. If someone threw down a stone from the Brahmaloka heaven at a certain time how long would it take to reach the earth, the abode of men? The answer was that it would take a whole year on its way through space.[7] We know too that the great Buddhist astronomer, I Hsing, in the early + eighth century calculated the time which had elapsed from the first General Conjunction as 96, 961, 740 years, which was a great deal more large-minded than the ideas of Western astronomers and theologians in the + eighteenth century.[8] So the Chinese astronomers, equipped with their celestial circles and free from the straitjacket of the Aristotelian-Ptolemaic celestial spheres, continued to think in terms of infinite empty space with bodies of unknown nature moving through it. It is paradoxical that the Chinese, who have so often been accused of being too materialistic and literal-minded, should have remained quite free from this concretisation of the starry heavens and the planetary spheres.

It is amusing that in the + sixteenth century when Matteo Ricci, the great Jesuit missionary, got to China, he wrote back in + 1595 telling of a number of 'absurdities' of the Chinese; one of which was that they said there was only one sky and not ten skies, that it was empty and not solid, and that the stars move in the void instead of being attached, as we know they are, to a firmament.[9] There is some reason for believing that later on knowledge of this *hsüan yeh* cosmology of China began to filter through to the West, and had some influence on the European astronomers when they were breaking through, or breaking away from, the fixed crystalline celestial spheres. For example, Christopher Scheiner, another great Jesuit astronomer, who disputed with Galileo the first discovery of the sun-spots, was writing in + 1625 to show that the realm of the stars had a 'fluid nature'. He remarked that

'the peoples of China have never taught in any of their innumerable and flourishing academies that the heavens are solid; or so we may conclude from their printed books dating from all times during the past two millennia. One can see therefore that the theory of a liquid heavens is really very ancient, and could easily be demonstrated; and moreover one must not despise the fact that it seems to have been given as a natural enlightenment to all peoples. The Chinese are so attached to it that they consider the contrary opinion [in other words, several crystalline solid celestial spheres] perfectly

absurd, as those inform us who have returned from among
them.'[10]

One could go on at further length about the astronomical
aspects of cosmology, but I would rather expatiate for the rest
of our time on the matter of the abodes of the dead. All civilisations have been deeply interested in this subject. If we leave on
one side the other-worldly realms of hells and paradises, ethic-
ally determined, and we talk only about the habitations of men's
spirits, more or less disembodied, in this present world, there
have been three possibilities: upon the earth, under the earth
in some subterranean realm, or somewhere in the starry heavens
above. The essential point is that if you exclude ethical deter-
minism, everybody ultimately went to one or other of these
universal and comprehensive places.

I append a diagram* which embodies these conceptions.
Some, like the Amerindians, have visualised 'happy hunting
grounds' where the spirits go. Others have thought that they
lived in their own tombs or barrows, or entered stones or trees.
Or they might be in some far away place on the earth, east or
west, or upon some island. It is of interest that many peoples
in different parts of the world have held similar beliefs. One of
the ideas most typical of Chinese thought, quite irrespective of
which cosmological theory was favoured, was that of the
possible continuance of the individual person with an etherealised body upon the earth: this type of existence could last
potentially through aeons of time, with the deceased paying
rare visits only to the habitations of men. More widespread
was the idea that everybody descended after death to an underground world of grey shadows. I suppose that this was the
dominant conception in ancient Israel, with the idea of She'ol,
and in Greece with Hades, but it is found in other civilisations
too.[11] The third idea, that of a realm of the dead situated somewhere in the heavens above, seems to have been less common;
but it existed in various places, and both the ancient Egyptians
and the primitive Patagonians believed that the souls of the

* See Figure 8, p. 94.

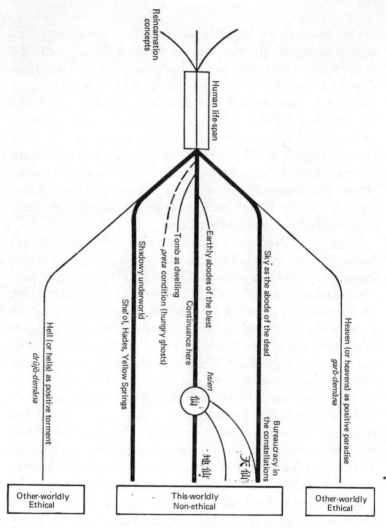

8. Schematisation of immortality conceptions in China; development of
ethical polarisation

dead went to find a home in the stars. That is important for us because Taoist religious philosophy in China also maintained that the perfected immortal could in time rise up to the constellations. These were part of the sky and lay within the confines of the natural world; but they formed the palaces and offices of the celestial bureaucracy, and there the distinguished immortal would find his appointed position.*

These ancient ideas of universal public places were invaded gradually by another conviction – that of a judgement which would separate good and evil men, rewarding the former and punishing the latter; this leads to what I call 'ethical polarisation', with a heaven and a hell outside this world, conceived in positive terms as paradise above and a place of permanent torment below. One source of this ethical polarisation might be ancient Egyptian, but another certainly seems to be Iranian, arising from Persian dualism and spreading outwards over all the parts of the Old World. In the Avestan sources there is talk of the *garō-demāna*, or 'House of Song' of Ahura-Mazda, standing over against the *drūjō demāna* or 'House of Lies' of Ahriman.[12] This is an Indo-Iranian idea and comes in the Vedas (by − 1000), where the good are supposed to ascend to the palaces of the gods. This ethical polarisation was taken over by everybody, or at least nearly everybody. Buddhism accepted it, with innumerable complications of hells, Christianity took it, so did Islam and most of the great traditions. In Israel, the old tradition of She'ol lasted down to about − 150. It is not until then that the idea of positive paradise and positive torment seems to come into Israel, so it was quite a late development at the time of the Gospels. Indigenous China never participated in this, however, and learnt of ethical polarisation only when Buddhism came flooding in, and that was not until the + third and + fourth centuries. Afterwards this doctrine exerted much influence, but never enough to overwhelm the ancient Taoist idea of a material immortality here on earth attainable by adepts and perfected spirits, who could thus avoid the common fate of descent to the underworld.

* See plate 18.

The indigenous Chinese position was fundamentally attached, then, to this-worldly levels, and there was a conception of a dark shadowy realm not unlike our own world, but underground somewhere in the neighbourhood of the Yellow Springs (*huang chhüan*), the name for the Chinese *She'ol* or Hades which became general early in the — first millennium. So if you were a *kai thien* supporter you might perhaps imagine that the Yellow Springs, where everybody went, were located far down underneath the lower dome. Of course if you were a *hun thien* supporter you would have located it inside the body of the egg-yolk or crossbow-bullet earth, surrounded far away outside by the celestial circles. It had to be inside, underground. But everybody went there, and servants and possessions were needed there, hence perhaps the human sacrifices at the royal burials of the Shang (— second millennium), and the wooden effigies like the Egyptian *ushabti* figures which you get in the Chinese tombs of the Chou (*ca* — 1122 to — 250) and Han (— 202 to + 220) periods; and the long tradition of sacrificial ancestor-worship too, which lasted unflinchingly, withstanding all argument, down to contemporary times.

It is important to remember that for the ancient Chinese conception a man was not a person with a single unitary soul; there was a group of two sorts of souls, some Yin and some Yang, partly ouranic and partly chthonic, so that if they were kept together at all they would have to be in connection with the body in some form or other. It might wander in the realm of the Yellow Springs, or on the surface of the earth, but if there was no bodily component at all the souls would dissipate into the world; some would go upwards and others would go downwards, to mingle with the *chhi* or vapour of the heavens and the juices (*i*) of earth. The body, however etherealised, was the string that strung them together.

Emphasis on longevity[13] kept growing in the early days of the Shang and Chou, and since being backed by one's ancestors and surrounded by one's descendants was the greatest blessing that heaven could confer, why should it not continue for ever? After the beginning of the — fourth century the conviction spread that there were *technical* means whereby man could enlarge his length of days so much as to be virtually immortal, not down in the Yellow Springs, not out of this natural world, not in any ethical paradise or hell, but among the mountains and forests, here and for ever. This was really the basis of the conception of a 'medicine of immortality', a strange part of Chinese cosmology indeed. The Chinese in this case really were very materialist; they believed that there were technical means whereby this persistence on the earth could be accomplished, running right along and not going up or down; neither descending to the shadowy underworld, nor floating about without any kind of patent of office among the stars, still less being consigned to eternal bliss or to everlasting damnation outside the natural world. The materialist aspect of it came in because the technical means of doing so was precisely the elaboration of a chemical drug or elixir of longevity and material immortality.

Something happened in the late Chou period, about the — fourth century, to strengthen this belief. It might have been a message from Babylonia, Persia or India about a drug-plant or a herb or some medicine of immortality, even perhaps slightly misunderstood, so as to interlock with Chinese world-views. The result was a great wave of activity concerned with what is sometimes called 'the cult of *hsien* immortality'. The word *hsien* means an immortal, living on or above the earth but within the world of Nature, a distinctively material immortality with a lightened body. The ancient Chinese believed in the existence of some drugs or chemicals or medicines which could be taken for this purpose. If one probes about for this in other civilisations one finds phrases which strike one's attention; for example among the Gnostics and the early Christians you find expressions like *pharmakon tés athanasias* (φάρμακον τῆς ἀθανασίας), the 'drug of immortality', but you find that they are always using it metaphorically. The Gnostics used it of baptism, Christians used it of the eucharist, but for the Chinese it was definitely 'tablets in a bottle'; if you could only find the right drug, the right chemical, you would be able to go on and on. All this certainly stands at the beginning of chemistry, just as some of the other ideas we have been speaking of lie at the beginning of astronomy. But a drug-plant or a mineral or metallic elixir is only naturally thinkable in the context of radical continuance. It could surely never be expected to guarantee the attainment of disembodied other-worldly bliss, or to protect against disembodied other-worldly torment, especially if both were considered deserved. It was essentially a medicine, and as medicines have to do with the maintenance or restoration of health here and now, so also an elixir had to do with the maintenance of health here and for ever. Our suggestion really is that all true alchemy in all civilisations was born from this specific situation and no other.

Maybe I should say something more about the Yellow Springs. If you ask any educated Chinese friend he immediately recalls the touching story of the ruling prince of Chêng State, whose name was Chuang, which occurs in the *Tso Chuan* under the date of — 721. His father had married a princess of Shen and of their two sons she preferred the younger one. But Chuang having ascended the throne his brother Tuan revolted and took many cities before being conquered. At one point in the family quarrel Tuan intended to besiege the Princess of Shen in the Chêng capital and she would have opened the gates to him but was prevented by loyalists from doing so.

'Chuang then shut up his mother in the city of Ying, and vowed and swore that he would never look upon her again until they both came to the Yellow Springs. Then after a time he repented of this oath. One Khao Shu, who was Warden of the Marches in the valley of Ying, heard of this and went to court to offer some present to the prince. The prince made him stay to dinner, and noticed that he laid aside some of the food, so he asked him why? "Your servant", answered Khao Shu, "still has a mother, who likes to taste of all the best things I am given to eat, and she has never tried this princely dish, so pray take no offence if I keep a little for her." The prince looked sad and presently said "Ah, you have a mother for whom you keep titbits! Alas, alas, I lack any parents, I have no mother". Khao Shu enquired how this was, and the prince told him the reason of it and his oath that he took after the civil war; whereupon Khao Shu remarked with a cheerful face, "Why should the prince grieve about this oath? If you would dig a tunnel in the earth going down to some springs, as is done for royal tombs, you could arrange to meet your mother down there, and who could then say that your oath had not been kept?" The prince followed this advice and entered the tunnel, chanting: "In this great subterranean place joy and concord will be found." His mother also came forth from it chanting: "Beyond the subterranean world our hearts are bursting with happiness." From that moment they resumed the relations of mother and son. Lords and masters in time to come will say that the filial piety of Khao Shu of Ying was perfect, for by his love for his own mother he awakened that of Prince Chuang of Chêng.'[14]

I think that this lovely passage does indicate the sort of idea that was held about the *huang chhüan*, a place not, after all, buried very deep below the earth's surface.

There are other passages, too, in which people talk about mines, and miners getting involved with the Yellow Springs, so they were not really very far down.[15] It must have been rather an alarming thought if you were mining for tin or copper to think that you might break down the wall and find yourself among the ghosts of the past in the land of the Yellow Springs. Wang Chhung, for example, about + 82, observed in fact that people naturally do not like the dark. Who would want to be a miner, he says, digging galleries in the vicinity of the Yellow Springs?[16] In any case, as a home of the dead, the idea went on continually through the history of Chinese thought and right down to comparatively modern times.

Another interesting story is one about the presentation of a 'drug of deathlessness'. The word deathlessness, *pu ssu*, occurs very often in place-names, country names, names of herbs and rivers and people and trees and drugs in ancient China. There is a story in the *Han Fei Tzu* book, written in the late — fourth century, about a man who taught the techniques of deathlessness to a Prince of Yen. What Han Fei says is that

'a certain travelling philosopher was once entertained at the Court of the Prince of Yen, and taught him something of the art of immortality. Afterwards the prince sent some of his young men to learn it more fully, but before they could complete their studies the philosopher died. The prince was extremely annoyed, and chastised the students. That prince never knew he had been deceived by the philosopher, and censured the young men for their dilatoriness, but to believe in an unattainable thing, and then to punish unculpable emissaries, is this not a calamity of unthinkingness? Besides, any man will have a care first and foremost for his own preservation, and if the philosopher was not able to make himself deathless how could he possibly make the prince live for ever?'[17]

This is the usual mixture of scepticism and sophistic argument, but what interests us is the fact that around — 320 there were men prepared to teach the art of achieving material immortality, and educated patricians who were eager to listen to them. It might have been physiological alchemy in this case, but it might also have been a matter of the ingestion of medicine, some drug-plant or some mineral chemical preparations, as well as the appropriate psychological exercises.

The next thing I would like to mention is that the *hsien* could be divided into two groups, the *thien hsien* or 'sky immortals', and the *ti hsien* or 'earth immortals'. The distinction between these is already present by the beginning of the + first century, and by the end of the Han, people were affirming the superiority of the *thien hsien*. Ko Hung (*c.* + 300), whom I mentioned before, wrote in his *Pao Phu Tzu* book, that great work on alchemy, that 'the manuals of the immortals tell us that masters of the highest category are able to raise themselves, their souls and bodies, high up into the airy void. These are called celestial immortals. Those of the second category resort to the famous mountains and forests and are called terrestial immortals. As for those of the third, they simply slough off the body after death and they are called corpse-free immortals.'[18] It does not explain where the last group went to reside, but presumably it was somewhere more agreeable than the Yellow Springs, maybe one of the less famous mountains or forests.

It seems that you could choose your aim, and in the *Shen Hsien Chuan*, a fascinating book of the early + fourth century concerning the immortals, we hear about Pai Shih *hsien-shêng* (Mr Whitestone), who chose the latter course. He was not going up into the skies – not he. Mr Whitestone was a disciple of Chung-Huang Chang-Jen, and in the time of Phêng Tsu (the Methuselah of China) he was already more than 2,000 years old. He was not willing to cultivate the *tao* of rising into the heavens as an immortal; he just wanted to be an immortal as such.

'He did not intend to do away with the joys and happinesses of life among men, so the course of action he adopted was to practice the arts of the bedchamber as the main thing [which is part of physiological macrobiotics] and emphasize the taking of the medicine of potable gold. In his youth he was poor and could not buy the drugs required, but lived as a shepherd and pig-keeper for more than 10 years, frugal of dress and diet. At length he acquired 10,000 pieces of gold, and was able to buy the great medicine and consume it. He often used to heat a certain white mineral with his food, and lived in the mountains near some white rocks, so people called him "Mr. Whitestone". On a day when he had eaten meat and drunk some wine he could travel three or four hundred *li*, appearing to those who met him not more than 40 years old at most. He loved the temple worship and the liturgies, and was fond of reading esoteric books. Phêng Tsu once asked him why he did not take the chemical which can make one rise into the heavens; to which he replied "Can the joys of the heavens really compare with those that are found among men? If one can go on living here below without getting old and dying, one will be treated with the greatest respect. Would one be treated any better in the heavens?" So the people all said that Mr. Whitestone was a *hsien* who wanted to avoid becoming a *hsien*. It was because he did not seek to rise into the heavens and to take a place among the celestial bureaucracy (*hsien kuan*). Nor did he have any desire for fame and renown in this present world.'[19]

There is a rather charming bucolic atmosphere in this story. It gives one the idea that the Han and Chin Taoists were really quite this-worldly, some of them appreciating the idea of 'eternal life in the midst of time', or, as some of our Buddhist friends might say, *saṃsāra* is *nirvāṇa*.

On the other hand there was a remarkable and widespread belief that certain immortals had been seen to ascend into the heavens in broad daylight, and Chinese books tell many stories of these ascensions. The *Hou Han Shu* (History of the Later Han Dynasty), an impeccable Confucian work, contains a circumstantial account of a performance of this kind by an adept named Shang-Chhang Kung witnessed by two well-known scholars who died around + 190.[20] We find also in the *Thai Phing Ching* (Canon of Great Peace and Equality), towards the end of the Han, just at the beginning of the + third century, a book which belongs to one of the Taoist sects or churches that were growing up at that time, a text which says that

'among the 36,000 things in the universe, longevity is the best of all. In this Heaven comes first, then earth, then the holy immortals in all of their eight ranks, who partake of the mind of royal Heaven and share its will and power. These are the kind of men that Heaven needs for official positions (in the celestial bureaucracy); so they all have concern for the same thing, that which Heaven most loves, the nourishing of the lives of men and women. Heaven most prizes longevity, beyond the span of ordinary life, and the immortals also most prize it; and those who prize such life dare not do the works of death, because each of them cares about the preservation of the bodies and souls of others as well as himself.'[21]

One can see here the appearance already at the end of the + second century of a distinct ethical element. Of course Confucian ideas of ethics will keep breaking in, as usual, but one need not complain of that.

Among the descriptions of ascensions into the heavens, the type specimen of all is Huang Ti himself, the Yellow Emperor. One can say that the prototype of all imperial flights and ascensions was that of the legendary emperor Huang Ti, and in — 113 Kungsun Chhing was telling about this event in considerable detail — it was supposed of course to have happened a couple of thousand years before.[22] After Huang Ti had cast a bronze tripod cauldron once upon a time at Shou-shan and brewed chemical elixirs in it, a celestial dragon vehicle came down from the heavens to fetch him, into the which he stepped, together with more than seventy other people, both ministers and palace ladies, and they all mounted up into the sky in full view of the populace. It is sad to recall that in some versions of the story certain lesser ministers held on by the hairs of the dragon's beard, but unfortunately these gave way so that they got left behind. Possibly that was a later anti-bureaucratic embellishment. This was the occasion about which it is recorded that a perfectly real and historical emperor, Han Wu Ti (Emperor Wu of the Han; reigned —141 to 87) afterwards unfeelingly said, 'Ah, if only I could become like the Yellow Emperor, I can see myself leaving behind my women and their children as lightly as casting off a sandal.'[23]

Now we are still talking about the central line in the diagram (Fig. 8), where the *hsien*, having taken the chemical elixirs or having performed the various activities of physiological alchemy, were rising as *thien hsien* into the heavens, not ethical positive other-worldly heavens, but the constellations of the sky as the seats of the heavenly bureaucracy. Perhaps I ought to interject that just as real Chinese society was bureaucratic through and through after the beginning of the Empire in the − third century for more than two millennia, so also it was very natural that a bureaucracy should be imagined in the heavens, ruling the world below, and this Taoist bureaucracy was obviously a reflection of the bureaucracy of earth. There are wonderful images and pictures showing the figures − for example the Director of Thunder in Spring, or the Assistant Head of the Smallpox Department in the Ministry of Disease, or again the Deputy Secretary-General of Cattle Plague, and beings of that kind, all along the line. So promotion into these ranks was dreamed of by the early adepts of the Taoist church in the + fourth and + fifth centuries; they had visions, they dreamed dreams, they believed that they would be called upon to rise into the heavenly bureaucracy in the stars above, not outside the natural world, but within it and within the constellations. There is a good illustration from a Han relief depicting the supreme ruler of the heavenly bureaucracy, Thai I, the Great Unity, or Supreme Unity (Head of the Civil Service), riding in the box of the Great Bear as if in a chariot. It comes from the Wu Liang tomb-shrines of + 147, and it illustrates the way in which the Han people thought of the gods and celestial bureaucrats among the stars (*see* plate 19).

One or two of these ascensions are quite extraordinary. For example, Liu An, the Prince of Huai-Nan, perhaps the greatest natural philosopher of the Han period, who died in − 122 or so, was responsible for getting a group together to compose the *Huai Nan Tzu* book, a great compendium of cosmography and proto-science. He was believed to have gone up too, having been a notable patron of alchemists, naturalists and magician-technicians. After supposedly planning sedition against the Emperor Han Wu Ti he was condemned to commit suicide in − 122, but after his death or disappearance (I always like to believe it was a disappearance) there quickly arose a rumour or legend that he had in fact ascended to the heavens as a *hsien*, and that not only with his whole family and household but also with all their domestic animals. All had ingested doses of a particularly potent elixir. There can be no doubt that much alchemical activity was going on at his court, and we even have the names of many of his adepts and laboratory operators, so the story may well have been put about by those who remained behind after the prince and his family had got away into those mountains and forests. In any case it was widely accepted in the folk traditions.

Another thing even more extraordinary is that Yen Kho-Chün preserved in his great collection of texts a remarkable Han inscription on stone entitled *Hsien Jen Thang Kung-Fang Pei*, 'Memorial of the Immortal Thang Kung-Fang.'[24] This tells us that in + 7, Wang Mang's time, Thang was a minor official in a district at Chhêng-Ku, a town in the Upper Han valley in the mountains dividing Shensi from Szechuan. By chance he happened to gain the friendship of a local adept, who accepted him as a disciple and gave him various chemical drugs, one of which made him understand perfectly the language of birds and beasts. Gradually Thang became a *hsien*, though continuing his employment; he could summon up any desired scenery in the neighbourhood for anyone who wished, and magically assembled and killed vast numbers of rats which had been devouring the bedding of the governor and the imperial envoy whom he was showing round the district. In spite of this however Thang Kung-Fang fell out with the governor, not being willing to teach him the Tao, so at last the governor ordered his underlings to arrest Thang and his family as well. Greatly alarmed, he sought the help of the adept, who duly proceeded to administer an elixir to Thang's wife and children, saying 'Now is the time to go'. But they were reluctant, poor things, to leave their home, so he asked them if they wished to take it all with them, and they said, yes, indeed, that was what they would like, so the adept daubed the whole house with a chemical preparation, giving an elixir also to the domestic animals. Whereupon there quickly arose a great wind and a cloud of darkness which completely carried away Thang Kung-Fang and his family and all their belongings. As the inscription goes on to say, this was much more extraordinary than the achievement of immortality by individual people like Wangtzu Chhiao and Chhih Sung Tzu. In fact one might consider it beyond the powers of modern chemistry also, were it not for the fact that nuclear physics and chemistry have in our own time been able to do the like; but alas it would not I fear be to any earthly paradise that the devilish mushroom-shaped cloud would sweep away Thang Kung-Fang and his family.

I should like to suggest, then, that in spite of metaphorical formulations, elixirs giving permanent life, or medicines of immortality, in other words what we like to call macrobiogens, were not a serious element in Hellenistic proto-chemistry, which concerned itself more with aurifiction and aurifaction. That is to say, gold-faking and gold-making – on the one hand imitations intended to deceive, on the other the belief that one had been able to synthesise gold from other substances. But the Chinese from the — fourth century onwards believed absolutely in these elixirs, and the kind of immortality they envisaged was fundamentally a physical or material one indicated on the diagram in Fig. 8 by the thick vertical line and its variants. It might be a perpetuation of human existence, the *hsien* immortal staying on in the light of the sun, not under the earth, or it might be that he or she would go up into the sky and reside among the constellations. There was no ethical polarisation until Buddhism came to China, and therefore a material medicine or elixir, of plant or mineral-metallic origin, was quite conceivable; for all medicines maintain or restore the health of the body-soul organism, and *hsien* immortality was after all only an indefinite continuation and etherealisation of this essential health. A material medicine could hardly be envisaged by anyone as a passport to any other-worldly ethically-determined paradise, or a safeguard against dismissal to a well-deserved purgatory or hell. The attainment of the status of the 'holy immortals', the *shen hsien*, depended primarily upon techniques, chemical, physiological, magical, not by all means necessarily ascetic in the usual sense, but a long training, whereby the body was etherealised and rarefied but conserved; that was the this-worldly immortality of China.

So these considerations, I think, go to show that from the — fourth century to the + first century intellectual conditions in China were extremely propitious for the development of the elixir idea. Where the concept of the 'drug of deathlessness' came from is another matter, but we are inclined to think that the basic idea originally came from other places, perhaps from India, even from Babylonia. We know about the Gilgamesh story, for example, about — 2,000. We know about Enkidu and the herb of immortality that he went to the bottom of the sea to get, and then unfortunately lost. There were many places it could come from, very ancient, I suppose as ancient as man himself. But the conditions were right in China just at this time for crystallising the idea of the elixir. And then if we follow the elixir idea downwards in time we find that although Hellenistic proto-chemistry did not have it, the Arabs undoubtedly did (after all *al-iksir* is an Arabic word). As soon as the + eighth century begins, then with Bālinās and the Jābirian corpus and later on al-Rāzī and al-Jildakī, and all the Arabic alchemists, all had got the idea of the elixir, as they said: 'the medicine of man and of metals'.[25] From that time onwards the road is very clear; one pictures the idea of the elixir coming westwards through the many Arabic-Chinese contacts, and then eventually passing on to the early Latin alchemists, who depended so much on the Arabs and translation from the Arabic. Coming through to Western Europe in the time of Roger Bacon and Albertus Magnus it goes right on along the road that led to Paracelsus about + 1500 with his wonderful slogan: 'The business of alchemy is not to make gold but to prepare medicines.' Modern chemistry, with Priestley and Lavoisier, may be said to have been born under the sign of chemo-therapy, and when one traces that back to its origins one finds them in the Taoism of ancient China.

Cosmology in that culture was thus associated both with astronomy and with chemistry as they developed through the history of the Old World. We have traced the ideas which the

Chinese formed of the visible heavens and the visible earth, that was one aspect of it; but also we have seen how their thought-system favoured the idea of human survival after death (or apparent death) in a glorious form, not in some other world outside space and time but in this one, here and for ever.

Notes

1. The conventions adopted in this chapter for dating, romanisation of Chinese names and terms, and references to Chinese texts, are those used in *Science and Civilisation in China* (hereinafter abbreviated as *SCC*). Further aspects of this subject are considered in greater detail in Vol. 3, 210 ff. and in Vol. 5, part 2 of that work.

2. The *li*, which is the regular Chinese unit for measuring distance, is generally regarded as equivalent to a third of a mile; but in contexts such as the present one it cannot be given a precise significance.

3. H. Chatley, 'The "Heavenly Cover", a Study in Ancient Chinese Astronomy', *Observatory*, 1938, **61**, 10.

4. For Professor Lambert's views see p. 62 above.

5. This rendering is based on a translation by H. Maspero, *L'Astronomie chinoise avant les Han'*, *T'oung Pao*, 1929, **26**, 267.

6 *Chin shu*, Ch. 11, 2a, cf. *SCC*, Vol. 3, 219.

7. Cf. *SCC*, Vol. 2, 420.

8. Cf. *SCC*, Vol. 3, 120, 408.

9. Cf. *SCC*, Vol. 3, 438–9.

10. *Rosa Ursina* (1630), 765; see *SCC*, Vol. 3, 442, note (a).

11. On such ideas see also Chapter 3.

12. Cf. L. C. Casartelli in *Encyclopaedia of Religion and Ethics*, Vol. 11, 847.

13. On this subject, see Yü Ying-shih, 'Life and Immortality in the Mind of Han China', *Harvard Journal of Asiatic Studies* (1964), 25, 80.

14. *Tso Chuan*, Duke Yin, 1st year, Couvreur tr., Vol. 1, 7 ff.

15. For example, *Mêng Tzu*, III, 2, x, 3, Legge trans. 161.

16. *Lun Hêng*, Ch. 38, Forke trans. Vol. 2, 99.

17. *Han Fei Tzu*, Ch. 32, p. 3a, Liao Wên-Kuei trans., Vol. 2, 39.

18. *Nei Phien*, Ch. 2, 9a, Ware trans., 47.

19. Ch. 2, biogr. 1.

20. Ch. 112B, 17b.
21. Wang Ming (ed.), 222–3.
22. *Shih Chi*, Ch. 28, 31a, b; Chavannes trans., Vol. 3, 488 ff.
23. *Chhien Han Shu*, Ch. 25A, 28a.
24. See Yen Kho-Chün, *Chhüan Han Wên*, Ch. 106, 1b–2a. So far as we know, there are no extant rubbings made from the original stele whereby one could check Yen's version, but there are no particular grounds for suspecting its genuineness.
25. Cf. R. P. Multhauf, *The Origins of Chemistry*, 117 ff.

5
Ancient Indian Cosmology

R. F. GOMBRICH
Lecturer in Sanskrit and Pali, University of Oxford

In 1835 Macaulay advocated English education as more likely
to benefit Indians than what they could learn from their own
traditional literature. Should the Government, he asked, 'countenance, at public expense . . . geography made up of seas of
treacle and seas of butter'? On this point at least we may
sympathise with Macaulay. Indian cosmology has not been a
popular subject, even with modern Indologists. It does not
lead anywhere. Its development had little or nothing to do
with the achievements of Indian science; even Indian astronomy
is conceptually separable from the rest of ancient Indian cosmology to a very large extent.[1] Moreover, it had no intimate
connection with those currents of Indian religion which have
attracted most attention in the West, and indeed been most
successful in the East: Buddhism, the idealistic monism called
advaita vedānta, and the devotional theism of *bhakti*. All these
three movements, or complexes of movements, are soteriologies
concerned with the individual's moral and metaphysical make-
up, and very largely indifferent to the physical universe outside
him. But the most discouraging feature of traditional Indian
cosmology is not its fantastic and uncritical character but its
complexity. Indeed, the title of this paper imposes the unenvi-
able task of outlining not one but four cosmologies. The earliest
known Indian cosmology, the Vedic, is very different from the

classical cosmologies which arose after the middle of the first millennium BC: the Hindu, the Buddhist and the Jain; and these three again, though not entirely unrelated, are quite distinct. Worse still, not one of these four major cosmological systems presents a single straightforward picture. Professor Kirfel's book *Die Kosmographie der Inder* (see bibliography) has over 400 large pages with hardly anything more than bare quotations and tables, and it covers only cosmography, the spatial arrangements of the universe. Forced to a choice, I have decided to devote three-quarters of this paper to Vedic and Hindu cosmology, not because Buddhists and Jains are any less interesting, but simply because the vast majority of Indians are and have been Hindus.

But first, why is Indian cosmology so complicated? Just as the Indian system of social organisation, caste, has grown throughout history by aggregation and inclusion, not abolishing the practices and customs of newly assimilated peoples but assigning them a low place in the social hierarchy, so Indian cosmology – which remained largely a branch of Indian mythology – rarely abandoned a theory or idea, but allowed it to remain alongside the new ideas, even if it was inconsistent with them. India is a very large and diverse area, which has been politically fragmented for most of its history, and Sanskritic civilisation took about two thousand years to spread over the whole of the sub-continent – indeed some peoples in remote areas remained little affected by it till the present century. Nevertheless, there are certain Hindu texts, the Purāṇas, composed since the beginning of our era, which concern themselves with five topics, of which two are the universe in space and time, that is cosmology; and the Purāṇas do make attempts to reconcile various versions and to present a systematised picture – though no two attempts give quite the same result. Systematisation proceeds, as I have just suggested, by aggregation and encapsulation; for instance, different cosmogonies are generally accommodated by making them occur successively, rather than by, say, interpreting one story as an allegorical alternative to another. It is this, I think, which largely accounts for the notorious fact that the dimensions of both space and time in the classical Indian cosmologies are so unconscionably large; two systems are reconciled by putting the one inside the other, and making it a cosmographical or temporal part of a much larger whole. We are about to meet a universe of Chinese boxes. I fear that this paper will mainly convey an impression of the vastness and complexity of my subject; but if I have an idea to offer in compensation, it is this simple one: that the enormity and complexity are like those of the enormous and complex social system, and exemplify the Indian tendency not to supersede new cultural elements but to juxtapose them with the old, in a hierarchic ranking.

To proceed, then, to facts. The *Ṛg Veda*, our earliest Indian document, dates from the second half of the second millennium BC. It contains two basic views about the construction of the universe: that it is bipartite, consisting of sky and earth; and that it is tripartite, consisting of earth, atmosphere and heaven. Kirfel thought that the former was the earlier, and he was probably right; certainly it was tripartition which became popular with later cosmologers. For already here the Chinese boxes start: sometimes the two, sky and earth, sometimes the three, earth, atmosphere and sky, are said each to consist of three strata, giving a total of six or nine.[2] When sky and earth are spoken of as the complementary pair they are called Dyaus and Pṛthivī respectively, Dyāvāpṛthivī (in the dual) together. Pṛthivī means 'the broad one (feminine)', and Dyaus is masculine. He is addressed in the vocative, 'O Father Heaven', as Dyaus Pitar, which is easy to recognise as Ζεῦ Πάτερ or Jupiter; Father Heaven and Mother Earth are Indo-European figures. Dyāvāpṛthivī play only a minor part in the *Ṛg Veda*: but they are mentioned as the parents of the world, being compared for instance to bull and cow.[3] However, they are also said to be created in various ways. But this paper cannot explore cosmogony as well, and must neglect it except when it has a direct bearing on cosmology in the narrower sense. Dyaus and Pṛthivī are compared to the two wheels at the ends of an axle, in which case the earth must be conceived as flat, but also to two bowls, and to two leather bags, in which case the earth is presumably concave. Kirfel interprets the two-bowl image as implying that the lower bowl is the underworld, with the earth as the diameter where the two bowls join, but evidence for his view seems weak. The cosmos consisting of earth and sky is compared to an edifice; besides building, the metaphors of weaving and sacrificing are used. The sky is said to be propped up, but there is also reference to the marvel of the unsupported sky. The earth is said to be fastened with bands or pegs.

If we turn now to the tri-partition, in this context earth, atmosphere and heaven are called *bhūr*, *bhuvah* and *svar* respectively, although some synonyms are also used. These three words, in the nominative as just cited, are from very early times known as the three *vyāhṛti*, the three utterances, and every orthodox brahmin has to pronounce them twice daily, preceded by the sacred syllable *Om*, as part of his twilight ritual (*Manu*, II, 78). In this ritual context the meaning of the syllables has long been disregarded; but *bhū* and *svar* are good classical Sanskrit words for earth and sky, and the middle term, *bhuvah*, is but the plural of *bhū*. This curious detail may suffice to show that it is the basic tri-partition which really pervades Indian cosmology. Moreover, the most ancient commentators of the *Ṛg Veda* of whom we have any knowledge already divided the gods into three classes, according to whether they inhabited earth, atmosphere, or sky.

I turn now to the sub-divisions, and move from the bottom up. The *Ṛg Veda* says that men inhabit the highest of the three earths; this would imply that there is an underworld, or rather that there are two underworlds; unfortunately there is nowhere in the *Ṛg Veda* any explicit mention of an underworld, though there seem to be hints. Normal candidates for inhabiting an underworld would be demons and dead sinners. All Indian cosmologies believe in various kinds of ghouls and sprites who live here on earth; Vedic Indians also had a class antithetical to the gods, the *asuras*. The *asuras*, like the Greek Titans, lost a war with the gods; they are a kind of fallen angels, though not so clearly evil. The *Ṛg Veda* associates the gods with light and the *asuras* with darkness; but only later texts specify that they live in darkness below the earth. Similarly, the *Ṛg Veda* consigns the wicked dead to literal or metaphorical obscurity: hell is not mentioned. There is indeed one prayer (*RV*, VII, 104, 11) that the sacrificer's enemy may lie below the three earths; but this should perhaps be compared to the prayer in the *Atharva Veda* (VI, 75, 3) that the enemy go beyond the three heavens: the idea may be just to get him right out of the universe.

The sub-division of the atmosphere, which is after all visible, was problematic. To quote A. B. Keith (pp. 5–6):

'In the atmosphere also there are three spaces, or often only two – one the heavenly and one the earthly – and in either case the highest is sometimes treated as if it were the heaven or sky itself. Like the earth it has rocks and mountains; streams (clouds) flow in it; and the water-dripping clouds are constantly compared to and identified with cows. It seems clear that the earthly as well as the heavenly portion of the atmosphere is above, not below, the earth, so that the sun does not return from west to east underneath the earth, but goes back by the way it came, turning its light side up to the sky and thus leaving earth in darkness.'

It is only the *Aitareya Brāhmaṇa*, a slightly later text, which says that the sun shines upwards at night; but the same theory may be implied by the *Ṛg Vedic* statement that the sun's steeds draw both light and dark light. As for the heaven, in its highest third reside the ancestral spirits (the *pitaras*, literally 'fathers') with their king Yama and the divine form of *soma*, the hallucinogenic drink at the centre of the cult which the *Ṛg Vedic* hymns celebrate.

Before leaving the *Ṛg Veda* I must refer to two of the many remarkable hymns which occur in its tenth and last book. Both hymns are primarily cosmogonic. In one (*RV*, X, 129) the poet speculates about the origin of the world:[4]

There was not the non-existent nor the existent then; there was not the air nor the heaven which is beyond. What did it contain? Where? In whose protection? Was there water, unfathomable, profound?

There was not death nor immortality then. There was not the beacon of night, nor of day. That one breathed, windless, by its power. Other than that there was not anything beyond.

Darkness was in the beginning hidden by darkness; indistinguishable, all this was water. That which, coming into being, was covered with the void, that One arose through the power of heat.

Desire in the beginning came upon that, desire that was the first seed of mind. Sages seeking in their hearts with wisdom found out the bond of the existent in the non-existent.

Whence this creation has arisen; whether he founded it or did not: he who in the highest heaven is its surveyor, he only knows, or else he knows not.

One idea in this hymn which becomes common in the later Vedic literature is that the world started somehow with a golden germ of fire which sprang up within the water. Fire and water, heat and moisture, sun and rain: however expressed, this particular polarity has been crucial in Indian mythological thought from the earliest times. Philosophy was soon to add earth, air, and (usually) ether as the other material elements; but for the interpretation of both the macrocosm and the microcosm the contrast and interplay of fire and water have been most important – perhaps not surprisingly, given the Indian climate.

The other cosmogonic hymn I must mention is the *Puruṣasūkta* (*RV*, X, 90), the *Hymn of the Cosmic Man*, Puruṣa; for it is the earliest document for many major themes in Indian civilisation. In this hymn the gods sacrifice a giant to create the physical universe:[5]

Puruṣa is this all, that has been and that will be. And he is the lord of immortality, which he grows beyond through food.

Such is his greatness, and more than that is Puruṣa. A fourth of him is all beings, three-fourths of him are what is immortal in heaven.

From his navel was produced the air; from his head the sky was evolved; from his feet the earth, from his ear the quarters: thus they fashioned the worlds.

The four estates of society, brahmins, etc., were produced from his mouth, arms, thighs and feet respectively. I introduce this hymn mainly for its explicit equation of the macrocosm and the microcosm, an equation crucial to the development of Upaniṣadic thought which culminated in 'Thou art that' – a dictum which equates the individual spirit with the world spirit.

I must deal much more briefly with the rest of Vedic literature, the sacred texts which were composed approximately in the first half of the first millennium BC. The fourth and last Veda, the *Atharva Veda*, does know of hell, and in the Brāhmaṇas, the class of texts which chronologically follows the Vedas, it becomes clear that the good go to heaven, the bad to hell. The universe is gradually being ethicised. One text describes a hell where men cut up and eat each other; another says that the animal you eat here will eat you there. Even now hell is not clearly located, but it is dark, and presumably underneath the earth or the world. However, these texts clearly state that the *asuras*, the anti-gods, have been banished to the underworld; or alternatively that the gods, whose direction is the north, have banished them to the south. The south becomes the horizontal equivalent to the underworld; so that by transference it also becomes the region of death, and Yama, king of the dead, becomes (and remains throughout Hindu history) the guardian of the southern direction, even though he is less associated with hell than with the blessed dead in heaven.

The cosmography of the Brāhmaṇas is no more consistent than that of the *Ṛg Veda*. There are still allusions to the world as bipartite: the world is a tortoise, its arched shell the heaven, its flat underside the earth. In the *Chāndogya Upaniṣad* – the Upaniṣads follow the Brahmaṇas – the world is compared to two halves of an eggshell, the heavens one of gold, the earth one of silver. This simile springs from cosmogonic myths in which the world is, or starts off as, an egg. After this period

Brahmāṇḍa, 'the egg of Brahman', is the standard Sanskrit expression for 'the universe'. The eggshells also remind us of the earlier comparison of Dyāvāpṛthivī, sky and earth, to two bowls. In the Brāhmaṇas we get the first attempts to estimate the world's size: the distance from earth to sky is a thousand days' journey by horse (*Aitareya Brāhmaṇa*, II, 17, 8), or, more modestly, the height of a thousand cows standing one on the other (*Pañcaviṃśa Brāhmaṇa*, XX, 1, 9).

The tripartite universe is also going strong. But there are two interesting variants. By one, the three worlds of *bhūr*, *bhuvaḥ* and *svar* are increased to seven: on top are added (going upwards) *mahar*, *janas*, *tapas* and *satyam*. These words mean 'might', 'people', 'penance' and 'truth' respectively. I suggest that here we may be dealing with an original metaphor; one can easily imagine saying that the practice of austerities takes man above the highest heaven; or that higher still than penance lies the world of truth. 'Might' and 'people' are more difficult; but such metaphors are common in this class of speculative religious literature, which gives symbolic interpretations of the sacrifice and all associated with it. Be that as it may, the division into seven planes was to become standard in Indian cosmology.

The other interesting variant is perhaps rather a variant on bi-partition, for it cuts the universe into four layers: the water above the sky, the sky, the earth, and the waters under the earth. This quadripartite cosmology did not find general acceptance, but deserves mention for the waters above and below the rest of the world. The idea that there is water above the sky, not merely in the atmosphere, occurs in the *Ṛg Veda*. The waters under the earth, which of course remind us of Thales, do not; but they are alluded to several times in later Vedic literature, and we shall meet them in the earliest Buddhist texts. Late Vedic texts also state that the earth is surrounded by water, and this idea may be implicit in some passages of the *Ṛg Veda*. Other passages in the *Ṛg Veda* speak of two oceans, east and west, and others of four, perhaps at the cardinal points.

Kizfel concluded that Vedic cosmology showed Babylonian influence. He laid stress on the waters over the sky, a far from

universal idea but found in Babylon as in India, and wrote
that the Babylonians too believed the earth to rest on and be
surrounded by water. He wrote that in Babylonian cosmology
too the number three was important at first, and was then
overlaid by the number seven – under the influence of the
seven planets. But we cannot with certainty date the consideration of the planets as a distinct group of *seven* to before the
eighth century in Babylon. The other similarities which he
alleged are chronologically even more dubious. Babylonian
influence there must have been; but I think not before the end
of the Vedic period; tri-partition calls for no particular explanation; and there is no evidence that the seven planets are known
to Vedic literature. Moreover, the seven heavens could well
have developed without them, for seven is a favourite Vedic
number: it occurs more often in the *Ṛg Veda* (on a simple
word count) than any other number between three and ten,
and there are seven heavenly rivers, seven horses of the sun, etc.
My guess, then, would be that external influence may account
for the major changes between Vedic cosmology and the cosmologies which followed; and this fits the received idea that
Indian trade with lands further west began in the middle of the
first millennium BC.

The classical cosmologies of Hinduism, Buddhism and Jainism
all arose after about 500 BC (the Buddha and Mahāvīra, the
founder of Jainism, died shortly after that date), though our
evidence for them is mostly very much later. In particular the
Jain evidence does not antedate the first century AD; but that
may be mainly because the earlier Jain texts were lost. In the
second half of the first millennium AD, the Purāṇic Hindu
cosmology was gradually superseded, in secular contexts, by
quite a different cosmology, under the influence of Greek
astronomy, and in modern times, of course, western cosmology
has penetrated similar educated circles; but by and large it is
the classical cosmologies which have permeated the culture and
are still accepted by the mass of the population. Hindu and
Jain cosmology have been virtually confined to India, and by

the same token have perhaps more in common than either has with Buddhist cosmology; Buddhism on the other hand penetrated virtually all points east. As a necessary economy, I shall ignore what used to be called northern Buddhism – the Buddhism of Nepal, Tibet, China, north Viet-Nam and Japan, and confine myself to Theravāda Buddhism, the school dominant in Ceylon and most of South-East Asia.

Although the three cosmologies – classical Hindu, Buddhist and Jain – differ, I can offer a few generalisations to help as guides through the morass. In general the universe is ethicised, so that with various important exceptions the good go up and the bad go down, the higher up you are the better, and so on. Basically tri-partition continues, but now with the human world, the world we can see, in the middle; the heavens are above it and the hells below. To us Europeans this scheme, unlike the Vedic one, is familiar, and so easily intelligible. All cosmologies also agree that there is a world mountain running through the middle, its centre at our level but its top and bottom reaching at least one heaven and hell; this *axis mundi* is called Mount Meru[6] in Sanskrit, Sineru or Sumeru in Pali, the classical language of Theravāda Buddhism. Mount Meru is so important that even the astronomers who accepted the Greek idea that the earth is a sphere in space tended to leave a golden Mount Meru with gods living on it at the North Pole;[7] at the South Pole the *asuras* dwell, at the submarine 'Mare's Head' fire.[8] The cosmologies also tend to agree that the world at our level contains oceans in concentric rings, usually seven of them – though our own position in all this varies greatly – and all agree that the edge of our level is ringed by a circular range of mountains; thus our level is a vast disc, with a mountain in the middle and a mountain range round the edge. For Buddhists and for some Hindus there are innumerable worlds, but they all repeat the pattern of ours. Above all, the three cosmologies share a predilection for vast figures to measure both cosmic space and cosmic time. The Vedas have apparently no particular theory about cosmic time. However, all three classical systems agree that the universe moves in vast cycles, though there is disagreement about whether these cycles go on for ever; and that we are living in a period of decline, during which human life is getting shorter and things generally are going from bad to worse; Hindus and Buddhists also believe that major periods end in the destruction of the universe by fire and/or flood (fire and water). The time schemes of the developed cosmologies are however all so vast that from the cosmic point of view our era of decline is trivial; innumerable better times are ahead. For all cosmologies the inhabitants of the universe, non-human as well as human, are mortal, and are reincarnated after death in some sort of relation to the moral quality of their previous acts, their *karma*, until they are released from this cycle by whatever liberation the particular soteriology recommends; for most Hindus and for all Jains those released stay in the universe, near or at the top, but in Buddhist belief they leave it entirely.

I begin now with the Hindu view of the world, what is often known as Purāṇic cosmology. The Purāṇas were mostly composed in the first millennium AD. However, most of the matter of Purāṇic cosmology occurs, at least in inchoate form, scattered through India's great epic, the *Mahābhārata*. The *Mahābhārata* was also compiled over a very long period, but generally ante-dates even the early Purāṇas; let me just say that the *Mahābhārata* material I am about to quote is probably BC. Another very influential cosmological text is the first chapter of the *Mānava Dharmaśāstra*, the Laws of Manu, the most famous and authoritative of the Hindu codes of law and conduct; this must date from fairly early in the first millennium AD.

I shall now present *in extenso* a passage from the *Mahābhārata*, as systematised by Professor Frauwallner, because it is a relatively early passage which gives a not untypical cosmogony combined with a basic cosmology of both space and time. This account occurs in the Inquiry of Śuka, a section in the twelfth book of the *Mahābhārata*, and it is closely related to that in the first book of Manu. It rests on the theory of the four ages of the world, the four *yuga*. The *yuga* are named after the four throws of the dice, 4, 3, 2, 1; in descending order they are called *kṛta*, *tretā*, *dvāpara* and *kali*. These four dicing terms are already referred to the four *yuga* by an ancient commentator on a Brāhmaṇa text, so this theory probably antedates even the *Mahābhārata*. The first, the *kṛta*, is the golden age, whereas we live in the *kali yuga*, which started with the great war which is the main theme of the *Mahābharata*; this date was later identified as 17 February, 3102 BC.[9] I cannot here describe the golden age when the bull of Dharma, cosmic rightness, stood on all four feet; and the *kali yuga* we know all too well; I must confine myself to temporal proportions and dimensions. The golden age lasts 4,000 years, with a dawn and twilight of 400 years each; during it, men live to an age of 400 years. The four *yuga* are related in the proportion 4:3:2:1, so that the maximum life-span is now 100 years, and the total length of our *kali yuga*, counting dawn and twilight, is 1,200 years. The four *yuga* together thus total 12,000 years; this period is called a *mahāyuga* (great *yuga*). This basic time-scheme of the four ages reminds us of Hesiod; and, if I may obtrude a personal detail, I distinctly remember being taught in primary school that there were four ages of man, old stone, new stone, bronze and iron, and that we live in the iron age. The only difference was that this was progress. We shall see, however, how this simple scheme of four declining *yuga* was varied and encapsulated in larger systems. Already in our *Mahābharata* passage it is vastly inflated, and it is easy to see why: once the beginning of the *kali yuga* was dated by astronomers at 3102 BC, simple arithmetic showed the epic bards that the *kali yuga* would already be over if it lasted only 1,200 years.

The Inquiry of Śuka explains the Indian computation of time, building it up from the smallest unit, the blink, which is about 1/5 of a second. I translate Frauwallner (p. 117): 'A day of 30 *muhūrta* [a *muhūrta*, the nearest unit to our hour, has 48 of our minutes, so 30 *muhūrta* = 24 hours] is a day for men. It consists of day and night. The day is for work, the night for rest. Just so a month is a day of the fathers (*pitaraḥ*), the spirits of the departed. The light half of the month . . . in which the moon waxes, is their day and serves them for activity. The dark half, when the moon wanes, is their night and serves for rest. Finally a year is a day of the gods. The passage north-wards (*uttarāyanam*), that is the half-year in which the sun moves northwards and the day lengthens, is their day. The passage southwards (*dakṣināyanam*) . . . is their night. Such days of the gods are the units according to which the ages of the world (*yuga*) are reckoned.' Obviously this immediately multiplies our 12,000 years for a *mahāyuga* by 360, the number of days in an Indian year at that period,[10] and makes the *kali yuga* alone last 1200 × 360, that is 432,000 human years, while the *mahāyuga* lasts 4,320,000 human years. To return to Frau-wallner (p. 118): 'A thousand such world ages [*mahāyuga*] are a day of Brahma [the world spirit – neuter] which thus lasts 12 million years.' (Frauwallner is using divine years; to get human years, multiply by 360). 'A night of Brahma is the same length. A day is introduced by a creation of the world, which perdures throughout the day, the eras constantly rotat-ing. The night begins with the destruction of the world, and while it lasts cosmic quietude reigns. One day and one night of Brahma constitute a world period (*kalpa*), with a total dura-tion of 24 million [divine] years. This is the highest unit of time . . . the creation and destruction of the world follow each other eternally.'

At this point we leave Frauwallner and the *Mahābharata* for a moment. We have here, combined with the theory of four declining ages, the theory of the *kalpa*, a period which starts with the creation and ends with the destruction of the world. The *mahāyuga* is simply made a sub-section of the *kalpa*. The Purāṇas also have another kind of vast period, the *manvantara*, or Interval of Manu. Manu is the primaeval (and eponymous) ancestor of man, the first law-giver; he has many epithets, and Jacobi plausibly suggested that these gave rise to a theory of many Manus – a common development in mythology, especi-ally in Indian mythology – which in turn led to a theory of many successive eras in which the various Manus could operate. Seventy-one *mahāyuga* make one *manvantara*, which sounds very odd till we realise that fourteen *manvantara* make one *kalpa* (994 *mahāyuga* and their dawns and twilights lasting six *mahāyuga*); we then realise that the number seventy-one was reached by dividing the 1,000 *mahāyuga* of a *kalpa* by fourteen, to accommodate fourteen Manus, and fiddling round with what was left over. The *manvantara* is thus a sub-division of a *kalpa*.[11] The Purāṇas also know of a larger unit than the *kalpa*, the *para*. The *kalpa*, we have seen, is the day of Brahma, who creates the universe each day at dawn. According to the *Viṣṇu Purāṇa* his life lasts for a *para*, which is 100 such years, and then that is the end. The world is finite in time. At the beginning of our present *kalpa* half the *para* had elapsed, that is, Brahma had his or its fiftieth birthday.

Though this is the majority view, one major school of Hindu philosophy, the *Mimāṃsā*, holds that the world is eternal. This school is atheist, and argues against the notion of a creator god. These positions it shares with the Jains (see below).

Let us return to the Inquiry of Śuka. Its cosmogony is largely expressed in the categories which became particularly associated with the classical school of philosophy called Sāṃkhya 'enumerative'. Sāṃkhya holds that all matter, which includes mental phenomena – everything indeed but the soul – evolves from an undifferentiated state, often called *avyaktam*; that there are five material elements: ether, air, fire, water and earth, which appear first in their most subtle, later in their grosser forms; and that throughout all this there operate three principles, which are intertwined like the strands of a rope. The three principles, which go back to the Upaniṣads and play a large part in the *Bhagavad Gītā*, are *sattva*, literally meaning 'goodness', or 'existence', which is also light, *rajas*, literally 'dust', which is passion, energy and activity, and *tamas*, literally 'darkness', which is also heaviness and inertia. This said, I translate Frauwallner (pp. 118–19):

'When the world-night comes to an end, Brahma wakes up and has the world come out of him. There arises from him first the great being (*mahad bhūtam* – neuter), which still counts as the undifferentiated (*avyaktam*). From the great being springs thought (*manas*), which already belongs to the realm of the differentiated (*vyaktam*). Thought is then the origin of the elements. From it comes ether, from ether wind, from wind fire, from fire water and from water earth. Each of these elements has its own characteristic property: ether has sound, wind has feel, fire visibility, water taste and earth smell. But they do not have these properties alone. Wind as well as feel has sound, fire as well as visibility has feel and sound, water as well as taste has visibility, feel and sound, and earth as well as smell has all the other properties . . . This concludes the creation of the underlying essences of which all things are constituted, and the creation of living creatures and worlds begins. First there arises the creator god Brahmā, alias Prajāpati. He creates the gods, the fathers, and mankind; also the worlds with all that fills them . . . finally he creates the Vedas and sacrifices, the orders of society and the stages of life.'

Here a theistic cosmogony has been added as a second stage to the more naturalistic model; the activity of the personified Brahmā, masculine, follows the events which proceed from the neuter Brahma, the world spirit, which or who is semi-personified only as it or he wakes from cosmic sleep. I may add that from the *Mbh.* on it is generally accepted that the personified Brahmā creates the universe with the cosmic principle of *rajas*; Viṣṇu preserves it with *sattva*; and Śiva destroys it with *tamas*.[12] The three principles of Sāṃkhya philosophy are thus paired off with three gods, whose appearance in this context as a Trinity so misled early European observers.

The text also describes the destruction of the world. Such destruction occurs at the end of a *kalpa*, but presumably was originally conceived as bringing the world to an end at the close of a *mahāyuga*, the smaller unit. Seven suns appear in the sky, and set the world on fire. Everything on earth is burnt, till the earth is as bare as the back of a tortoise. Now water re-assumes smell, the property of earth, and so on in the inverse order to that of creation. Finally the undifferentiated dissolves once more into Brahma, and Brahma alone remains.

How do the Hindus conceive of the construction of the

universe? The world is commonly called *tribhuvana*, the triple earth, and there is a basic tri-partition into the earth, the heavens above it, and the nether worlds below it. Consonant with the cosmogony just reported, though reflecting a more classical form of Sāṃkhya, there is a common theory that the world is shaped like an egg, the outermost shell of which is undifferentiated matter; within this is a layer of intelligence and within this egoicity; and within this again in due order are layers of ether, wind, fire and water. Each layer is ten times as thick as the next one in. The outermost layer of undifferentiated matter is sometimes omitted, but to include it gives the satisfactory number of seven layers. These layers envelop the entire universe; the water thus merges in cosmographic tradition with the water which in Vedic times was believed to be above the heavens and below and around the earth.

The earth, the element which comes into being last, does not encircle the universe, but forms a mass in the centre. Its basic shape might be described as a huge flat disc, as it was in the Vedas, but this disc is now broken up into a system of concentric oceans and continents. We – the cosmologers – live on Jambudvīpa, the rose-apple continent, so named after a giant rose-apple tree which stands to the south of Mount Meru. In some versions the name Jambudvīpa applies to all the land mass round Mount Meru, within the first ring of ocean, which is the ocean of salt; while in other versions Jambudvīpa denotes only the southern quarter of that land mass, namely India. In Buddhist literature Jambudvīpa definitely means India. But in Purāṇic cosmology the more normal nomenclature is for the whole middle continent, centring on Mount Meru, to be called Jambudvīpa, while the southern quarter of it, that is India, is called Bhāratavarṣa. (Bharat is the modern Hindi name and so the official name now of India.) Using this terminology, one can say that Jambudvīpa comprises four vast countries arranged round Mount Meru, one centred on each cardinal point of the compass. To the north of Meru are the Uttarakurus, reminiscent of the Hyperboreans; the lands to the east and west are commonly called East and West Videha.

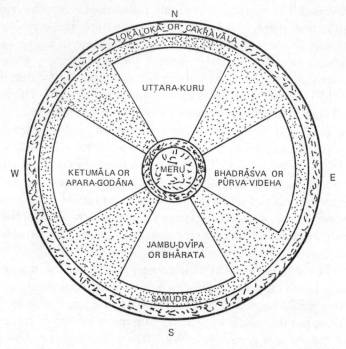

9. The Catur-dvīpā vasumatī. The earth consisting of four continents
(see pp. 136 and 142)[13]

Outside the ocean of salt are six other continents, all ring-
shaped and forming concentric circles round Mount Meru and
Jambudvīpa. Dimensions increase in geometrical progression
as one moves outwards. To quote L. D. Barnett (pp. 196-7)
(and to recall Macaulay), the continents

are divided one from another by ring-shaped oceans of sugar-
cane juice [which I take it is Macaulay's treacle], wine, clarified
butter, milk, whey (or whey and milk), and fresh water respectively. These circles are enclosed in a ring of mountains, called
Lokāloka, beyond which is a realm of darkness that extends to
the uttermost bounds of the universe ... The heavenly bodies
were supposed to have their orbits in planes parallel to that of

the earth . . . and to move round Meru at their centre. Their light is intercepted by Meru, and thus night and day arise. The variation in the height of the sun above the horizon is explained by the supposition that the sun's orbit round Meru varies, being narrowest at the summer solstice and widest at the winter solstice. The sun is imagined to move in summer more slowly by day than by night, and in winter more slowly by night than by day, the motions being equal only at the equinoxes.'

Mount Meru itself is generally said to be round and made of gold; it is 84,000 *yojanas* high and 84,000 *yojanas* deep, and it is the pleasure-ground of gods and demi-gods. There is a vast mass of mythical geography attached to it, as indeed to the rest of the world, but we may pass it over. The idea that a great mountain lies to the north of India is of course natural; the Himalayas themselves are generally accommodated as southern foothills to Mount Meru, with or without an intervening paradise. The Ganges drops from heaven – remember the Vedic waters of heaven – on to Śiva's head as he sits on Mount Kailāsa in the Himalayas, and thence flows down into India.

If the height and depth of Meru are 84,000 *yojanas*, the reader may wonder what a *yojana* is. It is generally translated as 'league', and unfortunately there are two classical Indian tables of measures of length: according to one the *yojana* would be about nine miles; according to the other, just half that. As a *yojana* is also used to denote a day's march the longer measure was probably the commoner; but in our context the matter is insoluble. The diameter of a *brahmāṇḍa* in Purāṇic cosmology is 500 million *yojanas* – the cosmic egg is apparently round, not ovoid; and I should perhaps mention that some texts would have it that there are an infinite number of them in space, though I suspect that this may be a late idea borrowed from the Buddhists.

For heavens, the Purāṇic cosmology generally retains the idea of the seven worlds which we found in the later Vedic literature. Above *bhūr*, that is us, *bhuvaḥ* extends as far as the

sun, *svar* from the sun to the pole star. The demarcation
between these three and the four super-added planes is still
clear, for it is only the bottom three levels that are destroyed at
the end of each *kalpa*. Of course this is inconsistent with the
idea of the seven envelopes round the whole universe and all
within them being created anew for each *kalpa*. The destruction of the world proceeds first by fire and then by flood. The
inhabitants of the *maharloka*, the fourth one up, though their
abode is not destroyed, apparently find it is too hot and wet for
comfort, for they move up to *Janaloka* for the time being.
Maharloka is ten million *yojanas* above the Pole Star, *Janaloka*
20 million above *Maharloka*, *Tapaloka* or *Tapoloka*, 80 million
above *Janaloka*, and *Satyaloka*, 120 million *yojanas* above
Tapoloka. *Satyaloka* is also called *Brahmaloka* and its inhabitants are truly immortal. They are the mythological equivalents
of those souls who have attained salvation by realising their
unity with Brahma, the world-spirit.

There are seven tiers of worlds underneath the earth; there
are many references to the universe as containing fourteen
worlds in all. These seven would correspond in a sense to the
seven upper worlds, from *bhūr* the earth to *satyaloka* the world
of truth; but the nether worlds are each only 10,000 *yojanas*
deep, a trifle compared to the heavens. Moreover there is great
confusion about the precise nature of these nether worlds, a
confusion which, as we saw above, already obtained in the
Vedic texts. There is an ancient and enduring class of demi-
gods important in the *Mahābhārata*, divine cobras called *nāgas*
who live at the bottom of river and seas, and/or underneath
the earth. These *nāgas* now join the archaic *asuras* and the
wicked dead as potential inhabitants of underworlds. Their
kind of nether world is called a *pātāla*. In Purāṇic cosmology
the seven levels below the earth are sometimes called *pātāla*
collectively, but they also have separate names, the lowest one
being Pātāla. The seven *pātālas* are splendid places, like
heavens; the lower ones are inhabited by *nāgas*, the higher ones
by various *asuras*. The *asuras* remain in most classical cosmologies as a Vedic survival, but the cosmologers are hard put

to it to fit them in. Below the bottom level, *Pātāla* proper, lies a particularly important *nāga* called Śeṣa, who plays a major part in Vaiṣṇavite cosmogony: he carries the world on his hood.

If the lower worlds are inhabited by semi-divine cobras, etc., where are the hells? Like the *pātālas*, they were presumably one at first, but have become seven, or multiples of seven – Manu gives twenty-one. As the addition of hells carries the total number of levels past fourteen, there is some doubt whether the hells come immediately below the *pātālas* or even lower down, outside the envelope of cosmic waters. This latter solution reminds us of the *Ṛg Vedic* prayer that one's enemy may lie below the three earths.[14]

The reader may be wondering where in the world one goes after death and how the heavens and hells fit in with reincarnation. When generalising I stated that all the inhabitants of the universe are reincarnated, and with the exception of the top heaven, *satyaloka*, that is true; but for Hinduism it is too simple a statement. When the doctrine of reincarnation appears in the Upaniṣads, the idea is that of those who do not have the liberating knowledge, the good are reborn on earth in a good station, presumably as well-off human beings, the bad as bugs, worms and so on. The idea of immediate rebirth on earth competes with that of reward in heaven or punishment in hell; and generally in the epics and Purāṇas we find both, so that the general picture is that a bad man suffers in hell and then is reborn on earth, e.g. as an outcaste. The good man may similarly go to heaven and then be reborn as a brahmin. But as ideas are hardly ever quite superseded, the Vedic king of the dead, Yama, must be fitted in even though the moral causation of *karma* is in principle automatic. In the *Ṛg Veda*, as we have seen, Yama and the blessed ancestors inhabit the top third of heaven. In the later Vedic literature there are also classes of ancestors who live in the earth, atmosphere and sky; besides which the souls of pious ancestors are sometimes said to be visible as the stars. Yama is the god of death, and he lives in heaven – the opposite of hell, according to the *Atharva Veda*; the two ideas,

I

his heavenly abode and his frightening character as Death, are not well integrated. In the *Mahābhārata* the good, especially good warriors, are said to go at death to Indra's heaven rather than to Yama's; Indra is the old warrior god, and his heaven is not firmly located, for the world of soteriology does not fit well with the world of systematic cosmology. Similarly, in late classical times the devout sectarian worshippers of Śiva and Viṣṇu believed that after death they might enjoy eternal bliss in the heaven of their god – but these heavens find no place in contemporary Purāṇic cosmology. But to return to Yama: he gradually loses his place as killer to a personification called Death, and retires to an apparent sinecure in his high heaven. This development is interestingly duplicated in Buddhism, where Yama's heaven is located immediately above Indra's but has no function. But there is an alternative theory that when the good die they go straight to the world of Indra, whereas the bad go to Yama, who condemns them to hell.

There are only about two million Jains in India now. But the Jains have devoted an extraordinary amount of attention to cosmography; I am told that it is a subject of absorbing interest even to contemporary laity, and that cosmographic diagrams appear in all Jain temples. This enables me to summarise a great deal by means of illustrations (see frontispiece and plates 21, 22). The universe is conceived of as a human being, either man or woman. Although the universe has a very odd – a unique – shape, I need hardly point out that this shape does not necessarily suggest the human figure; that image must rather derive from the tradition which we met in the Vedic hymn of the cosmic giant.

The Jain universe is measured in terms of a peculiar unit, the *rajju*, which is defined as the space covered by a god in six months if he flies at 2,057,152 *yojana* in a 'blink', that is about ten million *yojanas* a second. The measurements are all given in the most careful detail and the complications are enormous. The hells get wider towards the bottom, and there are vast inhabited gaps between them; the heavens, as you go up, get first wider and then narrower, and most of them are also bisected vertically. This bizarre structure must be the result of successive aggregations. We cannot unravel them all here; but Kirfel has found his favourite global shape encapsulated at the heart of this structure. Let me explain. The Jains, too, consider the terrestial disc to consist of concentric continents and oceans round Mount Meru, though they differ from the Hindus in detail; one peculiarity is that only two-and-a-half continents, counting out from the centre, are inhabited. Moreover the gods in the heavens live in vast celestial palaces, and the hells have similar sub-divisions, hells within hells as it were. Kirfel has pointed out that the bottom of the centre celestial palace of the lowest heaven precisely covers the two and a half continents of the inhabited terrestial plane, including the intervening seas of salt and black water, while the top precisely covers the central terrestial continent, Jambudvipā, and the top nether world corresponds, so that we can find a simple tripartite structure with the inhabited world at our level sandwiched between one heaven and one hell. Jain authorities themselves posit a kind of world within a world, though not in quite the same way as Kirfel; they say that although the entire world is packed with living things, those endowed with voluntary motion are confined to an area one *rajju* square which runs, like a funnel, right down the centre of the universe.

The Jains have no cosmogony; their doctrine denies a creator god. They hold time to be eternal. It is like a wheel with twelve spokes. The twelve spokes are twelve ages, divided into two sets of six: first there is the *avasarpini* or 'descending' series, the era of decline, then the *utsarpini* or 'ascending' series. This arrangement may be modelled on the twelve-month year divided into the *uttarāyana* when the sun goes north and the *daksināyana* when the sun goes south. At the same time, the ages are influenced by the four *yuga*, because if Jain ages 4, 5, and 6 in an era of decline are fused into one, we arrive at four ages with lengths in the proportion of 4:3:2:1. The dimensions dwarf even the Hindu estimates. We live in the fifth age of an era of decline. In the sixth age, at the end of the descending era, human stature finally descends to about a foot, the human life span to sixteen years, and men live brutishly in caves, feeding on raw flesh. As this is immediately followed by the ascending era, which starts with those same cosmic conditions, it follows that there can be no cosmic conflagration at the end of the era; but some Jain cosmologers do insert forty-nine days of corrosive rain (i.e. fiery water!) to mark the turning point, the nadir. Tīrthaṃkaras, the Jain saints who, like Buddhas, not only achieve enlightenment but broadcast the truth and establish it among men, are born, at least in one view, only in the third and fourth of the six ages. Buddhists hold a similar view. The idea is that in a golden age men do not feel the need of salvation, whereas in a really bad age the seed will fall on stony ground. The cycle of the six ages only takes place in Bhāratavarṣa, where we live; the other parts of the human world variously enjoy without change the conditions specified for one of the four better ages.

In contrast to Jaina time, Jaina space is strictly finite. Going from inside to out, the world is surrounded by layers of thick water, thick air, and thin air. Outside that is non-world, which is impenetrable, for it permits of neither rest nor motion, which in Jain philosophy are both substances. The top of the universe is interesting. Approaching the top we reach the gods 'of the neck' (the name refers to the cosmic giant), then the 'supreme' gods, who will be reborn not more than twice, and finally a comparatively small region, shaped like an open umbrella, in which liberated souls float in infinite knowledge and bliss just under the top edge of the universe.

A final bizarre detail: Jain astronomy posits that there are two suns, two moons, and two sets of the other heavenly bodies rotating round Mount Meru.

The cosmology of Theravāda Buddhism is fully evolved only in commentaries on the Pali Canon and in Buddhaghosa's *Visuddhimagga*, written in about 400 AD. A similar but rather more elaborate cosmology is to be found in the third chapter of Vasubandhu's *Abhidharmakośa* (fifth century). But the salient features of both are present piecemeal in the Canon, and have merely been systematised by commentators. Much of Buddhist cosmology thus has roots as old as the older parts of the *Mahābhārata*, and it developed fairly independently. Here we shall mainly follow Buddhaghosa.

For Buddhists, in contrast to Jains, cosmography is a secular subject in which interest should if anything be discouraged. For Buddhists, 'the world' can mean either 'the world of living beings' or the 'receptacle world',[15] i.e. space and time. Only the living beings are of religious concern, for the Buddha's message tells of the unsatisfactoriness of life, impermanent as it is, and of the way to escape rebirth by insight into the nature of things. Where other traditions talk of the world, therefore, in some such cosmological terms as 'heaven and earth', early Buddhist texts refer to the five *gati*, 'destinations', i.e. kinds of rebirth: gods, men, ghosts, animals and hell.[16] The first four of these terms refer to kinds of beings, of whom three – men, ghosts, and animals – are found on this earth; only hell is a locational term. It alludes primarily to the tortured sinner; but there are also demons inflicting the tortures, about whose cosmological position Buddhism has always been rather vague. We have here a three-tier universe with the earth between heaven and hell; its spatial and temporal dimensions are of no interest, and it is full of transmigrating beings who wish or should wish to escape it altogether.

Nevertheless, Buddhism soon constructed a fairly elaborate picture of the 'receptacle world', on the basis of hints and stray remarks in the Canon, and the reader will not be surprised that the universe of the five *gati* which I have just explained became incorporated in a much larger and more detailed scheme. But this incorporation took place in a very particular way. The universe was divided into three layers, called *kāmadhātu*, sphere of desire, *rūpadhātu*, sphere of form, and *arūpadhātu*, sphere of non-form. The world of the five destinations falls within – in fact it comprises – the bottom third, the sphere of desire. (The word which I have translated 'sphere', *dhātu*, is a term which only the Buddhists introduced into cosmology; its usual meaning is something like 'element', e.g. an essential ingredient of the body, or a grammatical root, and in classical Sanskrit has no spatial significance.) What we have here, I think, is a metaphor taken literally, the reification of a way of looking at spiritual progress. This becomes clearer when we look at details. At the top of the sphere of desire are the six heavens inhabited by gods; and in the highest of them dwells Māra, or Death, the personification who tempted the Buddha not to preach but to die, and whose daughters are the forces of desire, especially sexual desire. To put Māra in the highest of the six heavens gives the game away; it is a mythological restatement of the fundamental Buddhist message that just as Death and Desire are the two sides of one coin, so the entire world of transmigration, the sphere of desire, is presided over by Death. But Buddhist exegetes were puzzled that a palpably bad character like Māra should be in such a high heaven; hence the later idea that he lives in a remote part of this heaven with his hosts, like a rebel with his bands of brigands,[17] an apparent anomaly.

If the sphere of desire contains six heavens and all the old gods, one may well wonder what is in the two higher spheres, of form and no-form. The sphere of no-form, at the top, consists of levels with the same names – plane of neither perception nor non-perception, plane of nothingness, etc. – as denote various enstatic states which the Buddhist passes through on his way to the realisation of *nirvāṇa*. Clearly the heavens are not merely homologous with the states of yogic trance: they are reifications of the mystical terminology. They are inhabited by beings of pure mind, who died just short of the realisation of *nirvāṇa*, and are now not liable to a lower rebirth, but dwell in pure meditation which will ultimately be successful. The sphere of no-form, then, is merely an elaborate spatial metaphor for spiritual progress. Vasubandhu's *Abhidharmakośa* (2, 14, 4) is explicit that the sphere of no-form is non-spatial, and 'above' the sphere of form only in a figurative sense. However, the popular picture of the world is unlikely to be so sophisticated.

The sphere of form, between this and the sphere of desire, was a bit of a conundrum for cosmologers, who finally filled it with specially refined gods who derive from Mahā (great) Brahmā. We explained above that for the Hindus Brahmā, masculine, is the personal creator of the universe. The same god is often mentioned in the Buddhist Canon, a mere god but the highest one, who paid much honour to the Buddha. At first he was just the god of the highest heaven (so that with the six heavens of the world of desire there were in the Canon[18] seven heavens altogether); but then cosmologers lifted his heaven above the world of desire, and gradually elaborated and sub-divided it. It was not just that the sphere of form had to be filled: there was a more particular reason for putting Brahmā in it. Like Jains, Buddhists disbelieve in a creator god; but like the Hindus they believe that the world, or rather our part of it, is periodically destroyed. In the Canon[19] there is an amusing story at Brahmā's expense, a straight take-off of the Hindu view of personal creation: after the period of void (see below), Brahmā, because of previous *karma*, is the first being to be reborn; he is then lonely, and wishes for company, as in the *Bṛhadāraṇyaka Upaniṣad*; in the Upaniṣad other beings arise because of his desire, but for Buddhists this is a megalomanic delusion; the other beings too arise spontaneously because of their *karma*, but Brahmā fallaciously concludes 'post me propter me'. Thus he is outside the sphere of desire, but did not, as he believes, create it.

For Buddhists both space and time are infinite, although in the text just cited the Buddha reports both opinions but takes no position towards them beyond saying that his own insights are more valuable. There are infinitely many world systems, and post-canonical texts say that they are round spheres, which touch, and that the three-cornered interstices are special dark hells. This recalls the Hindu tendency to locate hell outside the bounded universe. However, the older idea was that the world (or a world) was bounded horizontally by a world-mountain with the relatively modest diameter of rather over a million *yojanas*, but the vertical dimension was not quantified,

presumably because of the unmeasurable nature of the sphere
of no-form. These two conceptions were combined by saying
that the Brahmā heavens, or some of them, extended over many
world-systems. In the Canon the earth rests on water, which
rests on air, which rests on space;[18] this is all later applied to a
world-system, so as to leave room for the hells under the earth.
For Buddhaghosa (*Vism.* XIII, 31) ten thousand such world-
spheres constitute a 'field of birth', and a hundred thousand
million constitute a 'field of command', over which the recita-
tion of Buddhist texts to avert evil is efficacious. It is a 'field of
command' which is destroyed at the end of an era (*kappa*).
There is also a 'field of scope', the scope of a Buddha's knowledge; that is infinite.

World geography at our level was unlike that of the Hindus
and Jains. Immediately round the cosmic axial mountain are
seven concentric rings of mountains, which originally perhaps
did not have seas between them; then between this central land
mass and the world-mountain at the edge is just one large sea,
in which four continents are situated as islands at the four
cardinal points, ours of course being the southern one. This
must be the kernel which has been encapsulated in a larger
scheme by the other traditions: the four-fold earth, a sym-
metrical scheme based on the four cardinal points of the compass, has been put inside a seven-fold earth. The seven-fold
earth has developed by taking the seven mountain rings which
encircle Mount Meru and interposing them between Mount
Meru and the mountain range round the edge, as seven circular
continents separated by seven seas. To keep the original four
continents, and to prevent the cosmographers' own country
from being consigned to some obscure peripheral point, the
four continents then became part of the central land mass, with
various mountain ranges, etc., intervening between them and
Mount Meru. All this is not pure conjecture, for schemes with
just the four continents occur in the older Purāṇas (*see* Fig. 9).
If the four continents are not islands but attached to one land
mass the problem of their boundaries arises; hence the Purāṇic
idea that they radiate from the centre like petals from a lotus.[20]

But that they did start off as islands is in my view proven, for the very word for continent in Sanskrit, *dvipa*, basically means 'island'.

'Buddhism knows neither a first cause of the world, nor an all-embracing spiritual substance giving rise to all that is. It is rather that something comes into being in dependence on and conditioned by something else. A first beginning is as impossible as is a definite end. The Buddhist, therefore, regards all attempts to explain the world or the individual by means of one or more 'eternal substances' (such as God, soul, original matter, atoms, etc.) just as useless . . . There are no permanent entities of any sort . . .'[21] It is this teaching of impermanence which for Buddhists is the *raison d'être* of their teachings concerning cosmic time. They combine in a complex and colourful way Jain-like ideas of ascending and descending eras with Hindu-like ideas of periodic conflagration and floods, followed by cosmic voids. Their scheme appears to systematise several ideas which appear in the Canon, where they are not coordinated. The main ideas may be listed:

1. The world periodically contracts (*saṃvaṭṭati*) and evolves (*vivaṭṭati*). (E.g. *Brahmajāla* and *Aggañña suttas*, *Digha Nikāya*.)

2. Time is divided into eons ((*mahā-*)*kappa*), each of which consists of four periods, called uncountables (*asaṃkheyya*):
   (i) the uncountable of contraction;
   (ii) the uncountable in a contracted state;
   (iii) the uncountable of evolution;
   (iv) the uncountable in an evolved state.

3. There are (unnamed) cycles in which the human life-span increases and decreases between the limits of eighty thousand and ten years, with corresponding improvement and deterioration in conditions. Decline is directly due to loss of virtue. At the nadir there is a period (*antara-kappa*) of a week during which people massacre each other or suffer some other disaster.

4. Even the sun and Mt. Sineru are impermanent: one day seven suns will appear and burn them up.

5. Even the four great elements – earth, water, fire and air – are impermanent. Of the last three it is said in identical terms that they may carry away anything from a village to a whole country; they may also be totally lacking.

Buddhaghosa's scheme (*Vism.* XIII) does not include the data from 3 (the *Cakkavattisīhanāda sutta, Digha Nikāya*); but he combines and reconciles the other ideas, plus a few minor points. The result is a scheme of an infinite series of great eons (*mahākappa*) which can be considered in sets of sixty-four. In these sixty-four great eons there is a cycle of types of destruction: seven by fire (by seven suns) are always followed by one by water; there are seven such octads; and then finally there are seven by fire followed by one by air. Destructions by fire, water and air follow the preponderance in the world of greed, hatred and delusion respectively. Each destruction (or 'contraction') destroys the 'field of command' up to a certain Brahmā world; water destroys higher than fire, and air than water. All beings are warned of each impending destruction, and this concentrates their minds to good effect, so that they are reborn high enough each time to escape the annihilation of their normal environment. The destruction of our physical world thus occurs at, or rather occasions, a high-point of religious progress; this odd doctrine was of course evolved to reconcile the destruction of the world with the indestructible chain of rebirth of unenlightened beings.

Each great eon begins with the uncountable of contraction, which itself starts with the appearance of a cloud of doom. There is then a great rain (which is welcome) followed by a great drought – fire and water again. Then follow either six more suns, or floods of caustic water, or a cataclysmic wind; these destroy the lower parts of the 'receptacle world' for the rest of the uncountable. The next uncountable is void. The next, that of evolution, begins with a great rain, which is compressed by air, and the world reappears from the top down. At our level there is at first only water, supported by air, but gradually scum appears on it and solidifies into earth. Beings die in the Brahmā Heaven of Radiance and are reborn in the sky; they are luminous, but on eating the scum lose their luminosity. Then the sun and moon appear, which marks the beginning of the fourth uncountable. From now on things decline and human institutions evolve; but this takes us from our subject.

Vasubandhu adds further elements. Each uncountable is sub-divided into twenty intermediate periods (*antaḥkalpa* = *antara-kappa*) and this leads to much scholastic elaboration. He works in the canonical idea of cycles of the human life-span (*see* 3. above), the Hindu theory of the four ages (*yuga*), and the Jain terminology of 'descending' and 'ascending' eras. Thus, at the start of the fourth uncountable the human expectation of life is 80,000 years. The first intermediate period is a descending one, and the life-span decreases to ten years. The next inter-mediate period is an ascending one; and this cycle is repeated nine times (giving ten cycles in all) to complete the uncount-able. A descending intermediate period is divided again into the four *yuga*, and the *kali yuga* begins when human life expectancy is 100 years. We are now in the *kali yuga* of the first descending intermediate period of our 'uncountable in an evolved state'.

But our situation is not altogether bad. At the beginning of each great eon, as many lotuses appear on the surface of the primordial waters as there will be Buddhas in the eon – any-thing from nil to five. Before our eon there was a run of twenty-nine 'empty' ones, with no Buddha. In our eon, however, there are five, the maximum, of whom Gotama Buddha was the fourth. There is one to come, and then – who knows? After Maitrī Buddha the next Buddha will be infinitely, un-thinkably, remote. We are very close to our last chance for salvation.

Notes

1. I would like to thank my friend Professor David Pingree for valuable comments on this paper. He writes: 'Indian astronomers were bothered by accusations that their acceptance of a Greek geocentric κόσμος contradicted *smṛti*; one will find attempts to justify their system in terms of Purāṇic concepts, especially in Brahmagupta's *Brāhmasphuṭa-siddhānta* XXI (AD 628) and in the eighth century commentator on this, Pṛthūdakasvāmin' (private communication).

2. For citations see Kirfel, pp. 4–5.

3. Contrary to my general proposition that concepts are not superseded, I should note that in classical times the idea of heaven and earth as the original parents become obsolete – so obsolete that in classical Sanskrit *dyaus* 'sky' is feminine.

4. Verses 1–4 and 7 are quoted in Macdonell's translation (1951).

5. Verses 2, 3 and 14 are quoted in Macdonell's translation (1951).

6. Mount Meru is mentioned in late Vedic literature – *Taittirīya Āraṇyaka* I, 7, 1 and 3.

7. With similar incongruity the Ionian Anaximenes, though he otherwise pictured the earth as a flat disc, seems to have posited a great mountain at the North Pole.

8. For much interesting information on this unusual combination of fire and water, see Wendy Doniger O'Flaherty, 'The Submarine Mare in the Mythology of Śiva', *Journal of the Royal Asiatic Society*, 1971, No. 1, 9–27.

9. Professor Pingree writes: 'The astronomical date for the beginning of Kaliyuga is 17/18 February 3101 (= 3102 BC); the date of the beginning of the Bharata war is generally placed later in astronomical texts. *Only* astronomers from the fifth century AD on and their imitators begin the Kaliyuga on 17/18 February 3101; others (including the author of *Mahābhārata* XII) give no specific date.'

10. '360 is the number of *saura* days in a sidereal year, where a *saura* day is defined as the time required for the mean Sun to traverse 1° of the sidereal ecliptic' (Pingree).

11. Astronomers, too, use the *manvantara*, but not always with 14 to a kalpa (Pingree).

12. It seems reasonable to assume that this idea is late in the *Mbh.*, and perhaps does not antedate the Christian era.

13. After D. C. Sircar (see Bibliographical note).

14. That there is a confusion between the *pātālas* and the hells was pointed out by Jacobi; he showed that the Jains have seven hells where

the Purāṇas put the seventeen *pātālas*, and accommodate the awkward archaic *asuras* in caves between the earth and the top hell – a much tidier solution.

15. This is the Sanskrit terminology. Cf. *Visuddhimagga*, VII, 37; in Theravāda Buddhism 'world' is said to have three meanings: all compounded things; living beings; and space (= our 'receptacle world').

16. Occasionally the awkward *asuras* form a sixth *gati*, between men and ghosts, but they are generally ignored in systematic cosmology.

17. See Nāṇamoli's note, *The Path of Purity*, 219.

18. For example, *Anguttara Nikāya*, i, 227.

19. *Brahmajāla Sutta, Dīgha Nikāya*, i, 17–18.

20. In Theravāda, *akāśa*, which is the fifth element 'ether' in most Indian schools, is not an element, but space.

21. This is the lotus which springs from Viṣṇu's navel; see Sircar, 264.

22. H. von Glasenapp, *Buddhism: a non-Theistic religion*, 50.

Bibliographical Note

The following bibliography may serve also as a list of suggestions for further reading.

I *Primary sources*
   Buddhaghosa, *The Path of Purification*, trans. Bhikkhu Ñāṇamoli, 2nd ed. (Colombo, 1964).
   G. Bühler (trans.), *The Laws of Manu* (*Sacred Books of the East*), vol. XXV (Oxford, 1886).
   A. A. Macdonell, *A Vedic Reader for Students* (reprinted Madras, 1951).
   L'Abhidharmakośa de Vasubandhu, trad. L. de la Vallée Poussin (Paris 1926); *troisième chapitre*.

II *Secondary sources*
   L. D. Barnett, *Antiquities of India* (London, 1913).
   E. Frauwallner, *Geschichte der Indischen Philosophie*, Vol. 1 (Salzburg, 1953).
   H. v. Glasenapp, *Der Jainismus* (Berlin, 1925).
   Articles in James Hastings, *Encyclopaedia of Religion and Ethics*:
   L. de la Vallée Poussin, 'Ages of the World (Buddhist)' and 'Cosmogony and Cosmology (Buddhist)'.
   H. Jacobi, 'Ages of the World (Indian)' and 'Cosmogony and Cosmology (Indian)'.

A. B. Keith, 'Indian Mythology' in *The Mythology of all Races* Vol. VI (Boston, 1917).

W. Kirfel, *Die Kosmographie der Inder* (Bonn & Leipzig, 1920).

A. A. Macdonell, *Vedic Mythology* (Strassburg, 1897).

D. C. Sircar, 'Cosmography and Geography in Early Indian Literature', *Indian Studies Past and Present*, VII (1965–6), pp. 233–334 and 353–407.

6
Islamic Cosmology

EDITH JACHIMOWICZ
University of Vienna

Western scholars have sometimes speculated whether or not Islamic ideas of cosmology have any claim to be treated as an independent branch of learning. It has been asserted, for example, that Islam is entirely indebted to Greece for its scientific knowledge of the cosmos, in so far as Islamic learning on the subject was derived from Greek ideas which found their way to the early Islamic scholastic centres in Damascus and Baghdad. At the same time it has been alleged that orthodox Islam has never hesitated to reject those doctrines that were incompatible with the spirit and law of the Koran, and that in consequence the cosmological ideas derived from Greece were to remain strange.[1] Such arguments appear difficult to refute, but they do not tell the whole story. They largely ignore the contribution which Islamic scholars have made to the ancient Greek scientific heritage and hence to the establishment of the modern view of the universe. They ignore, too, the fact that Greek cosmological conceptions were not so antagonistic to Islam as to make their integration into theological doctrine impossible.

Some pseudo-theological arguments are not strictly relevant to the issue at all. Their aim was to analyse the Koranic cosmos in such a way as to prove that it must have 'borrowed' much of its content from Jewish and Christian scriptural sources. We have here, therefore, an argument based on literary rather than doctrinal considerations. It has to be borne in mind that the

Koran, as the central core of Islamic religion, must be unique; to reject its distinct character would mean to reject the existence of the Islamic religion. Each religion must carry its own particular *Weltbild*, not least Islam. The religious cosmology of Islam is therefore inseparable from the revelation of the Koran and consequently is 'Islamic'.

With these short introductory remarks let us look now at the particular circumstances under which cosmological doctrine in Islam developed. It was determined on the one hand by the state of knowledge of the physical universe among the medieval Muslims. This was based mainly on Aristotelian and Ptolemaic theories filtered through Neoplatonism and Hermetism, and often enriched with Iranian gnostic elements. On the other hand, Koranic revelation and prophetic tradition presented distinct views about the cosmos and, even more important, provided the metaphysical basis for the integration into Islam of Greek doctrine about nature. Thus, by means of a synthesis of these two components, Greek heritage and Islamic revelation, Islamic cosmology was formed and elaborated into a system in its own right.

Perhaps because the quantitative outlook of modern scientific thought has tended to stress the physical component of cosmology more than the metaphysical, the Islamic contribution to cosmological science has been undervalued. So in order to study the proper status of Islamic cosmology it is advisable to reverse the usual order, that is to say, to consider first the Koran and the prophetic traditions (*hadīth*). This procedure is in fact more correct in both the logical and the historical sense, for it was not before the second century of Islam (eighth–ninth century AD) that the Muslims began to be acquainted with ancient Greek scientific and philosophical thought.[2] Let us therefore investigate the contribution of the Koran to cosmological thought, taking care to distinguish the metaphysical principles set in the Sacred Book from actual cosmographical facts.

The foremost principle has been given its most elaborate

and at the same time simplest expression in the first part of the *shahāda*, the Muslim proclamation of faith, *Lā ilāha illā Allāh*, 'there is no god but God'. Here we have in the Arabic language a clear expression of the absolute uniqueness of God, *aḥadiya*. This one God has created the heavens and the earth 'and all that is between them', a standing expression for the whole universe. Strict monotheism thus leads to another principle which was to become even more important in the development of cosmological doctrine in Islam: the principle of the essential unity, *waḥda*, of all existing beings through their common origin in a single source which is at the same time a common entelechy. We shall consider the full implications of this principle at a later stage, when we come to the metaphysics of Islamic cosmology.

A further important metaphysical principle laid down in the Koran is the special position it accords to man in the cosmos. Through his knowledge of all the Divine names, or aspects, regarded as the key to the knowledge of Nature, man has gained superiority over all other terrestrial beings and even in a certain sense over the angels. The second chapter (*sūra*) of the Koran contains the famous encounter between God and the angels concerning Adam, the first man:

'And when thy Lord said to the angels, "I am setting in the earth a viceroy". They said, "What, wilt Thou set therein one who will do corruption there, and shed blood, while we proclaim Thy praise and call Thee Holy?" He said, "Assuredly I know that which you know not." And He taught Adam the names, all of them; then He presented them unto the angels and said, "Now tell Me the names of these, if you speak truly". They said, "Glory be to Thee! We know not save what Thou hast taught us. Surely Thou art the All-knowing, the All-wise." He said, "Adam, tell them their names". And when he had told them their names He said, "Did I not tell you I know the unseen things of the heavens and earth? And I know what things you reveal, and what you were hiding." And when we said to the angels, "Bow yourselves to Adam"; so they bowed themselves, save Iblis; he refused, and waxed proud, and so he became one of the unbelievers'.[3]

The theories about the relation between the universe or macrocosm on the one hand and man or the microcosm on the other, as expounded by such eminent Islamic thinkers as the Ikhwān al-Ṣafā, Ibn Sīnā (Avicenna), 'Umar Khayyām, al-Ghazālī, and Ibn al-'Arabī,[4] to name but a few, can, it is true, be traced to Neo-Platonic sources.[5] The doctrines in their final form, however, supported as they were by passages from the Koran and the Ḥadīth, are certainly distinctively Islamic, since such an anthropocentric perspective is a characteristic feature of Islamic metaphysics.

To return to the Koran and its cosmological elements: the picture of the universe and its creation emerges from several scattered verses, but these rather obscure remarks together do not produce a very concise description of the cosmos. In fact, their ambiguity and occasional inconsistency gave rise to many different, even contradictory, interpretations, particularly with reference to the act of creation. Certainly, it was God who created the earth and the heavens; but whether this was a *creatio ex nihilo* or out of a primary matter, or how this primary matter came into being and what its nature was, were subject to numerous theological as well as philological disputes, in particular between the *Mutakallimūna* or apologetic theologians, and the *Falāsifa*, or peripatetic philosophers. We cannot dilate here on these problems since it would lead to a discussion of God's nature, the nature of beings, and the problem of time and causality.[6]

From the Koranic text it appears there that God created everything,[7] that in fact 'He is the first and the last';[8] that God's throne was above the water,[9] that heaven and earth were originally one solid mass, that God separated this mass, and that He created the heavens and the earth in six days[10] whereby the heavens were created out of smoke;[11] that He placed the mountains on the earth, and that He fixed the sun, the moon and the stars in the heavens, and created day and night;[12] that He created all living beings from water.[13] On the seventh day He seated Himself on His throne in order to rule over what He had created: 'He it is Who holds the heavens so that they do not fall on the earth',[14] and 'He it is who preserves heaven and earth from destruction'.[15] 'He directs the affair, He distinguishes the signs'.[16]

It is not difficult to imagine the vast possibilities which these verses offered to those thinkers who were seeking justification on Koranic grounds for their dialectical arguments. Thus it happened that the same verses were used by such orthodox theologians as al-Ash 'arī or al-Ghazālī to support their attacks upon the Mu 'tazilites, the followers of Greek philosophy, as well as by the Mu 'tazilites themselves and by the later mystics who were sometimes viewed with suspicion in orthodox circles. The importance of the Koranic text, not only with regard to all aspects of religious life, law and ethics, but also as a book of symbols, a cosmic cypher containing all the fundamental principles of existence, was realised and expounded by many Muslim authors. As a matter of fact, many of the esoteric practices of the Ṣūfīs and Dervishes employed cosmological symbols ultimately derived from The Book. It may be significant to observe that the Koran itself thus represents a perfect example for the second important principle mentioned above, that is unity, in a sense of the oneness comprising multiplicity.

Now as to the structure of the Universe, other passages in the Koran tell us of seven heavens and seven earths,[17] the latter being interpreted by Muslim cosmographers as the seven traditional climatic zones. In Muslim eschatology, however, these seven earths were identified with the seven strata into which the body of the world is horizontally divided and which are described as the seven mansions of Hell.[18] Above the heavens are situated the pedestal (kursiyy) and the throne ('arsh) of God.[19] At the demarcation between the seven astronomical spheres, which are those of the planets, and the superior spheres, at the gateway of the heavenly Paradise[20] grows a large tree known as the Lotus-tree of the Boundary.[21] Four rivers spring from its roots; two of them water Paradise, the other two are the earthly Euphrates and Nile.

These heavenly spheres and landscapes are populated by hosts of angels, the highest amongst them being the Cherubim. Islamic angelology, firmly rooted in the Koran, was later to develop under Iranian influence into an important branch of esoteric disciplines, with a strong impact on cosmological doctrine.[22] But already early Islamic tradition and legend based on episodes in the Koran are rich in descriptions of the world of angels. Collection and compilation of these texts was started at a very early stage, within a century of Muḥammad's death in 642 AD. They reflect the exuberantly colourful imagination of the pious Muslim. A brief discussion of one example which is the most significant as regards its cosmological contents will illustrate this.

Mention is made in the Koran of the Prophet Muḥammad's night journey (*isrā'*) from Mecca to Jerusalem and his subsequent ascent (*mi'rādj*) under the guidance of the archangel Gabriel through the heavens to the Throne of God, where the mysteries of Divinity were revealed to him. This story is only briefly referred to in the Koran in a single verse.[23] But pious Islamic imagination has elaborated this verse into long narratives containing a wealth of popular beliefs about cosmology and geography. In later centuries, particularly in Persian and Turkish literature, narratives and poetic cycles on the *mi'rādj* became a kind of literary genre, the *mi'rādjnāma*. In this process the heavenly regions of the Islamic cosmos were furnished with new details, partly of Manichaean and Central Asiatic origin, of which one of the finest examples is undoubtedly the *Mi'rādjnāma* of the illustrious Persian poet Abd al-Raḥmān Djāmī which is contained in the first book of his work *Silsilat al-Dhahab*.

The original story of the *mi'rādj* has come down to us in three main versions, two early and one later, probably of Persian origin. A full critical discussion of them has been given by Miguel Asìn Palacios.[24] What interests us here are the descriptions of the various stages of this journey through the cosmos. Here astronomy and popular Muslim eschatology are firmly linked together.

Ten is the usual figure given for the stages of the ascent. The first seven stages correspond to the seven astronomical heavens. At the gate of each the travellers, Muḥammad and the archangel Gabriel, are met by a guardian angel who welcomes them after having learned who they are. In each heaven Muḥammad is greeted by one or two prophets dwelling in that particular sphere. The most frequent order in which they appear is Adam, Jesus and John, Joseph, Idrīs (Enoch), Aaron, Moses, and Abraham. The prophet Abraham is seen leaning against the wall of the temple of the celestial Jerusalem, a replica of the earthly city. Then begins the ascent through the last three heavens. The first is represented by the gigantic 'Lotus-tree of the Boundary', with leaves as large as the ears of an elephant and fruits like jars, the second by the 'Inhabited Place', a Koranic expression for the temple of the celestial Jerusalem,[25] and the third and last by the Throne of God.

There are, however, certain variations in the descriptions of the different stages. In one version, the vision of a gigantic cock is introduced into the first heaven, an element of distinctly Persian origin. The wings of this cock stretch across the horizon and its crest touches the throne of God. The third heaven, with a fiery angel as doorkeeper, sometimes contains a vision of Hell.[26] There are seven levels in Hell, with fiery landscapes, cities and seas. The tortures inflicted in each of them are described with the exaggerated enthusiasm of popular religious imagination.

We similarly find descriptions of the heavenly Paradise inserted after the stage of the 'Lotus-tree of the Boundary'. Here also are seven subdivisions, with mountains and valleys and seas: the seas of light, darkness, fire, water, pearls, and snow. Unlike those versions that do not contain a full description of Paradise and that describe the theological heavens i.e. the last three, as being completely devoid of angels or human beings, these accounts of Paradise are regularly interrupted with chapters on the angelic populations dwelling in them, praising their indescribable beauties and splendours in words which stretch the expressive power of Oriental languages to the limit. In later versions we find celestial animals also introduced into the legend, such as the heavenly serpent that curls around the throne of God. Another later introduction is that of the heavenly beast, *Burāk*, which carries the prophet Muḥammad on his ascent. This feature must have been particularly popular, as we find it depicted on most of the miniature paintings of the ascent.[27]

Looking back on the structure of the legend, with its seven astronomical heavens, its seven levels of Hell, its seven mansions of the celestial Paradise, and its three stages of the theological heaven, we will recognise their correspondence with certain verses in the Koran, for example: 'Seven are the astronomical heavens and seven the earths, as are seven the seas, the gates of hell and the mansions of Paradise',[28] and other verses referring to the Lotus-tree, the Pedestal and the Throne.[29]

We can foresee the lines on which a legend so rich in vision was certain to develop. This was a shift to a purely allegorical meaning, the ascent being applied to other realities, spiritual or physical, and the images being taken to be mere symbols for intellectual experiences. The natural universe was now conceived as a simulacrum, a reflection of the hierarchy of being emanating from and reverting to its First Principle. Amongst the many authors of such spiritual *mi'rādjnāmas* we may find famous names like Ibn Sīnā, Suhrawardī of Aleppo, Ibn al-'Arabī, and al-Ma 'arrī.[30] We will return to this point later in connection with the cosmic function of man.

We have so far investigated the cosmological principles in the Koran, the few but significant details referring to the cosmogony and the structure of the universe, with examples of their use in Muslim eschatology. It remains for us now to show how the cosmos, as viewed by Islamic astronomical science, itself based on the Greek heritage, was able to preserve its identity as Islamic and Koranic.

This cosmos was conceived as a spherical entity, comparable in structure to an onion, with the earth at the centre. The part nearest to the centre is the sublunary region, which consists of the surface of the earth and the zone above it up to the lower boundary of the sphere of the moon. This region is referred to as the world of generation and corruption, the house of the 'Three Kingdoms' (*malakūt*)of natural objects, minerals, plants and animals. These are composed, in varying degrees of perfection, of the four elements, fire, air, water and earth,[31] and contain two of the four fundamental qualities, heat, cold, dryness and moisture. The four elements were often termed 'mothers' (*ummahāt*) by Muslim cosmologists. Those scholars who were following the Aristotelian Peripatetic doctrines, al-Kindī, al-Bīrūnī and Ibn Rushd (Averroes), for example, held that only the sublunary region was composed of the four elements; the celestial spheres and bodies were of a different substance, known as ether (*athīr*), and not subject to generation and corruption. Al-Bīrūnī furthermore maintained that the substance of the spheres was crystalline.[32] Despite the difference in substance assumed by the Peripatetics, both the heavenly spheres and the sublunary region were believed to possess the four fundamental qualities.

Other scholars, more inclined towards Hermetic tradition, believed the heavenly spheres to be composed of the same four elements as the sublunary region. Suhrawardī of Aleppo, for example, went so far as to abandon the Aristotelian distinction between the sublunary world and the celestial regions, proposing instead a new boundary at the sphere of the fixed stars, at the point where the region of pure light is divided from the world where light is mixed with darkness, or matter, in varying degrees.[33]

Whatever the connections were, their impact on the development of Islamic scientific astrology was enormous. The relation between microcosm and macrocosm was one of the subjects dealt with by Islamic philosophers and scientists, and astrology was only one aspect of it.

The sphere nearest to the earth is that of the moon. Because of its special position, the lunar sphere was seen as a kind of intermediary between the sublunary region and the celestial spheres, exerting an important influence on the world of

10. The natural location of the four elements in the sublunary region. In the centre lies the globe of the element earth, because heavy masses gravitate towards the centre. On the surface of the earth lies water, less heavy than earth. Surrounding the globe and extending to the sphere of the moon is the element air, which, by its friction with the movement of the lunar sphere produces fire. This is less concentrated round the poles owing to the slower movement there. (Reproduced from *The Book of Instruction in the Elements of the Art of Astrology*, by al-Bīrūnī, British Museum, Ms. Or. 8349.)

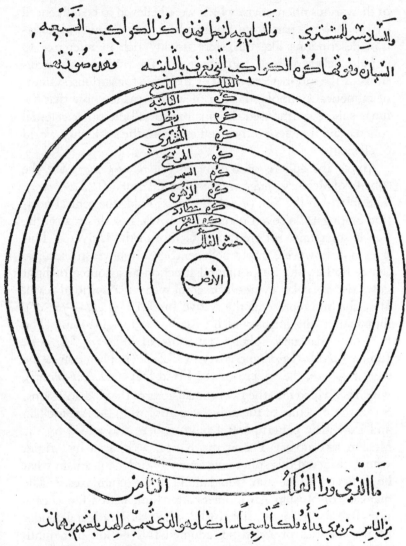

11. The structure of the Universe. Basically this follows Ptolemy, with the addition of the ninth sphere (from British Museum Ms. Or. 8349.)

elements and natural phenomena. We may quote al-Ḳazwīnī's *Cosmography* here. This treatise contains a detailed description of the various phenomena which were believed to be influenced by the moon. Apart from features based on scientific observation descriptions like this were usually rich in reference to popular beliefs as well.[34] Furthermore, the twenty-eight mansions of the moon were taken as the basis for various systems of numerical symbols, the aim of which was to show that the lunar sphere synthesised the qualities of all the other celestial spheres and transferred their influences to the sublunary world of elements.[35]

Above the sphere of the moon (*falak al-ḳamar*) are the spheres of Mercury (*falak al-'aṭārid*) and Venus (*falak al-ẓuhra*). The sphere of the Sun (*falak al-shams*) holds the central position amongst the seven planetary spheres. For this reason the Sun is sometimes referred to as Centre of the Universe. This does not, however, imply heliocentricity in any astronomical sense. Rather, the Sun's natural function as a source of light gave rise to solar symbolism employed at a spiritual level. Here the Sun represents the Active Intellect (*al-akl al-fa "āl*), illuminating the forms which exist in the imagination.[36] The last three planetary spheres are those of Mars (*al-Mirrīkh*), Jupiter (*al-Mushtarī*) and Saturn (*al-Zuhāl*). Then comes the eighth sphere, the sphere of the Fixed Stars (*falak al-burūdj*), which is divided into the twelve mansions of the Zodiacal signs. So far the structure of the universe conforms to the Aristotelian and Ptolemaic pattern, but a ninth sphere was added by the Muslim astronomer Thābiṭ ibn Ḳurra, a Sabaean by origin, who considered this sphere necessary in order to explain what he believed to be the trepidation of the equinoxes.[37] The majority of Muslim astronomers after him, with a few exceptions like al-Bīrūnī, and Ibn Rushd, who was an ardent defender of the theories of Aristotle, adopted this notion of a ninth sphere. It was usually referred to as the Sphere of Spheres (*falak al-aflāk*), or Englobing Sphere (*al-falak al-muḥīṭ*), and was believed to be starless.

The mutual influences of the heavenly spheres was the

subject of various studies undertaken by Muslim astronomers, physicists and mathematicians. Whatever their observations, it must be repeated that the ultimate purpose of these treatises was to study these phenomena in terms of their relation to their primary source. A quotation from the *Taṣawwurāt* of the great Persian astronomer Naṣīr al-Dīn Ṭūsī reveals this attitude in discussing

'the force of the creative act which, through the process of creation, reached the Throne of God, from the Throne reached the Pedestal, and, from the Pedestal again, descended to the sphere of Saturn and became attached to it. Again, it descended further, from one sphere to the other, until it reached the sphere of the Moon. Then the exhalations and rays of the stars, by the force of that energy and through the mediation of the sphere of the Moon, fell upon the elements. This was certainly the cause which stirred the elements (that they would begin to mix with each other).'[38]

In Greek cosmology the movements of the planets and the spheres were usually thought to be circular.[39] Naturally, the results achieved through individual observation often contradicted this theory, and explanations had to be found in order to 'save the phenomena'. Thus, in his *Syntaxis* which was to become famous in the Latin Middle Ages under its Arabic name, *Almagest* (*Kitāb al-Madjisti*), Ptolemy introduced the notion of the epicycles and eccentrics of the planets. This was to account for the irregularities of the routes of the planets within their spheres. Arab scientists eagerly adopted this theory, and proceeded to develop and elaborate it. Al-Farghanī, al-Battanī, and Abdurrahmān al-Ṣūfī, to name only a few, made valiant efforts to correct the theories in the Almagest regarding spherical motions and the distances and measurements of the planets. Ibn al-Haitham composed his famous treatise 'On the Structure of the Universe' (*Fi hay'at al-'Alam*), which was strongly influenced by the pseudo-Ptolemaic treatise *Hypotheseis*.[40] He studied the motions of each planet, with its eccentrics and epicycles, but with the Aristotelian assumption of homocentric orbs. As a physicist he concerned himself at length with the question of the substances of the spheres. The theory of solid spheres, known to the Greeks before Ptolemy, was adopted by Ibn al-Haitham mainly as a result of his belief in the physical impossibility of a vacuum. On the other hand, scholars like al-Farghānī and al-Battānī, who were astronomers and mathematicians rather than physicists, regarded the spheres as merely mathematical entities.

Generally speaking, the aims and criticisms of Muslim astronomers were mainly directed towards the improvement and further elaboration of the Ptolemaic system. In fact, however, these endeavours led to the final destruction of the system and, in conjunction with the general dissatisfaction felt about it, paved the way for the Copernican revolution.

The most rigorous criticism of Ptolemy was brought by the Islamic Peripatetics, who sought to revive the Aristotelian theory of homocentric spheres which Ptolemy had superseded. In Spain, the Islamic West in particular, a scholastic tradition of criticism of the Almagest arose, starting with Ibn Badjdja (Avempace).[41] Under his influence, the astronomer Djābir ibn Aflākh composed the famous treatise on 'The Rectification of the Almagest' (Iṣlāḥ al-Madjisti). A little later the philosopher Ibn Ṭufayl (Abubacer) encouraged his disciple al-Biṭrūdjī (Alpetragius) in his criticism, which was then expounded in the polemical 'Treatise on Astronomy' (Kitāb al-Hay' a). This revived the theory of homocentric spheres and at the same time introduced a theory of the spiral motion of the spheres and of the trepidation of the equinoxes. This, already anticipated by Thābit ibn Ḳurra three centuries before, was essentially wrong. Nevertheless, al-Biṭrūdjī's merit must not be undervalued since it was his ideas which had brought a heavy attack against the Ptolemaic system. In the Latin Middle Ages al-Biṭrūdjī's doctrines were praised as a new astronomy and thus were without doubt indirectly responsible for later developments.

We may add that although the cosmic system as conceived

by Islamic scientists was based on the assumption of the circular motion of the spheres, the possibility of a motion other than circular was considered. Al-Bīrūnī, in his correspondence with Ibn Sīnā, discussed on one occasion the Aristotelian thesis that an elliptical or lentil-shaped revolution needs a vacuum in order to make its movement possible while a circular revolution requires no such vacuum. As far as the physical aspect of this thesis was concerned, al-Bīrūnī did not object. However, from a strictly logical point of view an elliptical or lentil-shaped revolution appeared to al-Bīrūnī perfectly possible without need of a vacuum.[42] He continued: 'I am not saying that I believe the shape of the great heavens is not spherical but elliptical or lentil-shaped. I have made numerous studies to reject this view, but I do wonder at the logicians (that they were unaware of the possibility of such a motion)!'[43]

One thing has emerged clearly from the preceding paragraphs: the main anxiety felt by Muslim astronomers about Ptolemaic theories were with regard to the motion of the spheres and the form of the planetary motions within their spheres. Improved methods of scientific observation and new discoveries, together with the high standards of Arabic trigonometry, fostered the objections made against the Almagest. But the real reason for the dilemma, the essentially erroneous conception of the basic structure of the universe itself, was to remain virtually uncorrected by Muslim scientists. Although, as we have seen, many of Ptolemy's theories were at times strongly criticised, some of his doctrines being proved false, others being corrected or improved, and still others being felt intuitively to be wrong without a better solution being proposed, the general pattern of the geocentric universe was more or less accepted by Islamic astronomers.

Various opinions have quite often been formed in retrospect to account for the apparent stagnation in Islamic scientific thought about the cosmos. The cultural decline of the Islamic world during the political disturbances from the thirteenth century onwards, the growing unpopularity of Peripatetic

rational philosophy in conjunction with the increasing inclina-
tion towards mysticism, an exuberant religiosity, a swing from
exoteric to esoteric speculation, a triumph of *ḥikma* (sapientia)
over *'ilm* (scientia) – these were some of the factors which have
been held responsible for the failure of the Muslims to change
the medieval picture of the universe. Certainly these factors
have contributed in one way or another. But the real problem
is not simply a historical one, nor is it religious. The doctrines
of Islam and the cosmographical features in the Koran would
have produced no essential obstacle to the concept of a heliocentric system, as these principles were metaphysical, and the
given facts were vague enough to allow for more than one
interpretation.

A quotation from al-Bīrūnī, who was probably the most
gifted natural scientist of the Islamic Middle Ages, may help to
throw some light on this question. When examining an astrolabe constructed by the astronomer Abū Sa 'īd Sidjzī and
based on the heliocentric system, al-Bīrūnī remarked on 'the
idea entertained by some that the motion we see is due to the
earth's movement and not to that of the sky. By my life it is a
problem difficult to solve or refute . . . It is the same whether
you assume that the earth is in motion or the sky. In neither
case does it affect the astronomical sciences. It is for the physicist
to see if it is possible to refute it.'[44] This means that al-Bīrūni
was fully aware of the fact that heliocentricity and its implications were a matter of physics, and that, in order to assess its
significance for astronomical theory, the physical possibilities
had to be investigated first. After life-long studies, al-Bīrūnī
himself came to a negative conclusion as to the possibility of
heliocentricity: he was not able to verify its physics. A few
other Islamic scholars, notably Ibn Rushd, toyed with the idea
as well, but it invariably remained for them merely a hypothesis. It would be rash to conclude from this that Islamic
physicists were deficient in ability, and indeed any historical
account of the achievements of Muslim scholars in the natural
sciences would seem to reduce any such idea to absurdity.

What thus appears to present a discrepancy may perhaps be

clarified by examining the Islamic study of physics in its relation to metaphysics. For Muslim scholars the realm of physical
bodies, their qualities and behaviour, has always been secondary to and derivative from the plane of archetypes, or Platonic
ideas reflecting the Divine qualities. In other words, physical
phenomena were not studied as objects in their own right, but
only in their function as corporal, quantitative offshoots of
qualitative realities. A glance at any of the Arabic scientific
compendia reveals the same pattern. Metaphysics comes first,
serving both as point of departure and of return, or, in Arabic
terms, *mabdā* and *ma ʿād*. Physical objects are the last link in
the ontological chain of being, the plane of greatest variety
and multiplicity. Thus the aim of Muslim physicists in studying
natural phenomena was to establish the link between them and
their archetypes in order to trace them back to their common
single source.

The question of heliocentrism or geocentrism thus became a
matter of intellectual speculation, but it was never in Islam one
of fundamental importance with regard to theories about the
universe. Hence we cannot attribute the failure of Muslim
physicists to solve the problem of heliocentricity to any lack
of intellect or creative imagination. It was a matter of perspective, of what was considered to be important. In any case,
although the Muslims certainly made remarkable discoveries
in the field of physics, it was in the field of the Quadrivium,
the sciences deriving from mathematics, that Islamic scholar-
ship proved itself most brilliant.

We must therefore glance briefly at the Muslim attitude
towards mathematics, of which astronomy was a branch and
which therefore affected Islamic cosmological doctrines. The
science of numbers and geometrical figures was studied by the
Muslims in the Pythagorean manner, that is to say in their
qualitative and symbolic, as well as their quantitative and
numerical, aspects. This means that each number and each
geometrical figure was considered in relation to its metaphysical basis. The various series of numbers starting from the
source, which is the number one, the variety of geometrical

Plates

18. Taoist-Buddhist woodblock peasant print of comparatively recent date, showing the celestial bureaucracy mirroring that on earth. In the top row the three universal sages, Lao Tzu, the Buddha and Confucius. Below, following the centre line, the Jade Emperor first person of the Taoist Trinity, the Head of the Civil Service, the Goddess of Thai-Shan, and the King of the Underworld. Note at the top, right and left, the gods of the Northern and other Dippers; and along the bottom row the Directors of Horse and Cattle Plague, the City-God and the Genius loci. Left centrally one finds the Director-General of Floods and Droughts, right the Superintendent in charge of Landslides and Earthquakes. Many other officials of lower rank, some concerned with lucky and unlucky days, fill up the rest of the space. (Original in the East Asian History of Science Library, Cambridge, kindly presented by Dr Hsio-Yen Shih, Toronto.

19. The Great Bear (Northern Dipper) carrying one of the celestial bureaucrats; a relief from the Wu Liang tomb shrines (ca. + 147) reproduced in *Chin Shih So*, Shih sect., Ch. 3, hence *Science and Civilisation in China*, Vol. 3, Fig. 90, p. 241. The high official, doubtless once himself a Taoist adept on earth, and later one of the holy immortals, is met by minor officials at some point on his daily journey round the pole-star.

20. The ascension of Liu An, Prince of Huai-Nan, patron of a group of alchemists and natural philosophers (−122). From *Lieh Hsien Chhüan Chuan*, Ch. 2, p. 25a.

21. A diagram of the Jain universe. Paintings from Rajasthan (western India), + or − eighteenth century (from A. Mookerjee, *Tantra Art* (New York and Paris, 1966).

22. Another diagram of the Jain universe (from P. Rawson, *The Art of Tantra* (London, 1973).

23. Muḥammad carried by the Archangel Gabriel over the heavenly landscapes. Mountains and seas of fire, populated by angels. Miniature in Mongol-Persian style, attributed to the Master Aḥmad Mūsā, from a Mū 'rādjnāma Ms., dated second quarter of the fourteenth century, Topkapi Sarayi Müzesi, Ms. Hazine 2154, fol. 61a.

24. *Above:* Muḥammad and Gabriel are met by a Guardian Angel at the gate of one heaven. *Below:* The two contemplating the Tree of Paradise. From the same Saray Album, fol. 42b.

25. Muḥammad mounted upon Burák. From the same Saray Album, fol. 68b.

26. Stone from Sanda. A large stone (340 cm. high) discovered in pieces under the floor of Sanda church in Gotland in 1954. The stone was originally painted and belongs in a series of memorial stones with whirling discs and ships with rowers, but the tree and the dragon head mark it out from the rest. It probably dates for the beginning of the sixth century.

27. Larbro Tängelgårde stone. Memorial stone from Larbro Tängelgårde, Gotland (No. 1), of Viking Age date. In the top panel is a battle scene, with eagles hovering above. Below is an eight-legged horse, which seems to be carrying the dead man, and below this a rider is welcomed into the realm of Odin by a woman bearing a horn and men holding rings. The ship of the dead, seen below this, and the welcoming scene are found on many Gotland

19

20

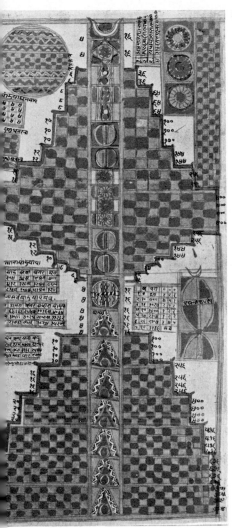

26

27

28

29

30A

30B

31

stones, but the bearing away of the dead man on Odin's horse after battle is
an unusual feature (from H. E. Davidson, *Pagan Scandinavia* (London, 1967).
28. Gylfi consults the gods. An illustration from an early manuscript of
Snorri Sturluson's *Prose Edda*, a handbook of mythology written about 1220
in Icelandic. It shows King Gylfi seeking information about the old gods from
three mysterious figures seated one above the other, who are called High,
Just-As-High and Third, three names attributed to the god Odin.
29. Stave church, Borgund. A wooden 'stave' church of twelfth-century date
at Fagusnes, Borgund, in Norway. The construction of these churches, of
which about 180 still stood at the beginning of the nineteenth century, is
based on wooden pillars resting on a 'sill' of beams and the series of roofs and
the dragon heads are characteristic.
30 (a). Thorwald's cross-slab, Andreas (No. 128). Outstanding tenth-century
cross-slab, although incomplete. Graphic scene from Ragnarok, the last
great battle of Norse pagan mythology. Odin (recognised by his raven on his
shoulder, and wielding his famous spear) is devoured by the Fenris wolf. The
wolf itself is later slain, in common with everything else ('and every living
thing shall suffer death . . . and the Powers shall perish').
(b) The opposite face of this cross shows a belted figure, bearing a book and a
cross, trampling on a serpent. Alongside is a fish (an early symbol of Christianity). This carving seems placed in significant juxtaposition to the scene of the
end of the world of the pagan gods. Christ now reigns in Odin's stead.
31. Tynwald Hill. A drawing of Tynwald Hill, Isle of Man, made about 1774,
probably by Godfrey. Here the Manx Parliament has assembled since the
time of the Viking kingdom of Man. The representative of the reigning
monarch now sits on the summit with the Bishop, the two Deemsters on the
level below, and below again the members of the House of Keys (from H. E.
Davidson, *Pagan Scandinavia* (London, 1967).

figures built up from the point, whereby their essential connection with the number one was always maintained, offered a perfect symbolic system for the understanding of the metaphysical principle of unity in multiplicity. This principle was also a cosmic one. A few quotations from some of the most eminent Islamic thinkers will illustrate this. According to the *Ikhwān al-Ṣafā*, the science of numbers is 'the first support of the Soul by the Intellect, and the generous effusion of the Intellect upon the Soul.'[45] 'Umar Khayyām stated that 'the number necessarily embraces everything'.[46] Particularly relevant is al-Fārābī's remark that 'it is characteristic of this science of numbers and magnitudes that its directives are identical with the principles of being.'[47]

This passage is important in so far as the principles of being are the object of the discussion that follows. Cosmology is necessarily closely linked with ontology, and ontology in Islam is concerned with the study of the hierarchy of beings descending or emanating from the one Being. In order to make this chain of beings in its relation to Being intelligible, Islamic scholars established systems of symbols mainly drawn from the science of numbers and from cosmology. Thus seen, the cosmos of corporal forms became the stage, or the surface of a mirror where the permanent archetypes, the ideas behind all natural objects, were present, or reflected, in their variety of forms, spiritual or corporal. These archetypes, however, are ultimately seen as aspects of the one Being. This Being has been referred to by various names, such as God, the First Principle, the First Cause, or the Prime Mover; and, depending on the philosophical system, one of these expressions was used for denoting Being in its most abstract sense.

Earlier in this paper we dealt with the act of creation, with the historical event of the generation of the Universe. From an ontological point of view, this universe was generated through Divine self-manifestation by means of a gradual concretisation and differentiation of Being through several stages, the lowest[48] of which was the realm of natural objects composed of the four elements. This process was frequently referred to by Arab writers as *tadjalli*, a term usually translated 'emanation'. The basic idea of this was that Being, in its highest and most abstract sense, is completely void of any attributes and qualities; it can only be defined in terms of itself: it *is*. In order to make intelligible the essential relation between this abstract Divine Essence and the multiplicity of beings in their infinite variety of forms – a relation which can be ultimately realised as an essential unity – a system was elaborated of descending stages of existence, or 'presences' (*ḥaḍarāt*) of Being. At each stage Being manifests itself in a certain degree of differentiation; it is 'present', that is to say, in the particular forms which constitute the hierarchy of beings.

Muslim writers usually postulated five, sometimes six, stages of Being. The first and highest is the stage of the Divine Essence or *ḥaḍrat al-dhāt*, Being in its unspecified oneness and absolute transcendence. This stage is therefore also named *Hāhūt*, which means Selfhood. In more esoteric language, we may further come across the name World of the Absolute Mystery, *'ālam al-ghayb al-muṭlak*. From this stage the process of the self-manifestation of the Divine essence starts, first reaching the second stage which is called Presence of Divinity (*ḥaḍrat al-ulūhiyya*), and represents the stage of Divine Names, that is to say the attributes and qualities of Divinity. This stage is sometimes also referred to as *lāhūt*, meaning that at this level of existence the Divine essence manifests itself through its own qualities. All the intelligible beings which constitute the remaining stages of the hierarchy are nothing but derivatives of these archetypal Divine names, in essence undifferentiated from them. This is why the second stage became associated with the Universal Intellect, *al-aḳl al-kulliy*. From this the individual intelligence emanate, thus forming the third stage, that of the *djabarūt*. These intelligences are the direct agents, the executive forces of the Divine command, and this stage is therefore referred to as Presence of the Masterhood, *ḥaḍrat al-rubūbiyya*. The angelic substances are usually identified with the intelligences, which signifies the ontological position of the angels.

The fourth stage is established as an intermediary – or isthmus – (*barzakh*) between the pure intelligences of the angelic world and the realm of material objects. The soul and its faculty of producing images (*mithāl*) through imagination (*khayāl*) constitute this stage of existence. It is characterised by individual forms and figures of subtle, incorruptible matter. Here are reflected, as in a mirror, the realities which belong to the higher planes of existence, appearing in shapes similar to those of the material bodies. Since they do not possess corporality however, they cannot be perceived through the natural senses. They can only be experienced through that particular 'sense' of the soul, the imagination which forms the link between the spiritual and physical worlds. This realisation of the ultimately symbolic nature of natural forms in all their infinite varieties, and the discovery of their inner meaning makes it possible for the human mind to understand realities which are completely beyond physical existence. This plane of existence greatly interested Islamic mystical philosophers, particularly from Suhrawardī onwards.

The fifth and final plane is that of sensible experience, *mushāhada*, comprising the world of composite bodies including mankind, *nāsūt*. The prototype of man, known as the First Man, *al-insān al-awwal*, or Perfect Man, *al-insān al-kāmil*, or simply Adam, used to be given a separate and superior position in the hierarchy of beings. This was the sixth plane, mentioned above, in which all other planes of existence were thought to be contained. This idea of a human ontological prototype can be traced back without difficulty to Neo-Platonic sources, in particular to the Hermetic texts and to the so-called Theology of Aristotle, a summary of Plotinus's Sixth Ennead. In Islam, however, from al-Ghazālī onwards, and later particularly in the Shi 'ite gnosis, Perfect Man emerged as a cosmic principle of paramount importance. In his function as a microcosm he was seen as the inner aspect of the whole creation, unity and variety combined in one being, and therefore the most noble and perfect representative of the Highest Being Itself.

After this first consideration of the structure of existence, with its various ontological strata, it remains to study the means by which this system of existences actually worked. This may be described as a continuous process of intellection in conjunction with an interplay of active and passive capacities. It would take a whole book to do justice to all the Muslim writers who have discussed this subject, but I may confine myself here to one scholar of the later Islamic Middle Ages: Naṣīr al-Dīn Ṭūsī, the philosopher, scientist and founder of the so-called Maragha School,[49] Islamic cosmologist *par excellence*, of particular significance for the final trends in Islamic philosophical and scientific thought. The battle between Peripatetic and non-Peripatetic doctrines, which had raged during the centuries before him, was finally moving towards the victory of the latter, and Naṣīr al-Dīn was one of the foremost figures in this tendency. Al-Fārābī, Ibn Sīnā, 'Umar Khayyām, al-Ghazalī, Ibn al-'Arabī, and many others, all of them dealt with the process of intellection fully in their writings. This was the background against which Naṣīr al-Dīn worked. Consequently, in his treatise *al-Taṣawwurāt* we find this subject expounded with particular clarity.

This is the scheme: God, Who is the source of all existence, became conscious of Himself as a Creator. The Divine will to create necessarily demanded the existence of a created object. This created object had to be a unified entity, as it was created directly, and without intermediary, by the Divine will and according to the principle that one can only produce one. This creation was the First Intellect, *al-akl al-awwal*, a unified whole which conceived the ideas of all spiritual and material beings below it. Three modes of cognition were effected by the First Intellect, thus bringing about three other kinds of existence. By reflection on its own cause – that is, the Divine will – the First Intellect brought about the existence of the Second Intellect, which is associated with the sphere of the Fixed Stars. Next, the reflection of the First Intellect on its own substance, as being necessary with regard to God's consciousness as a creator, produced the Universal Soul, which is the soul of that sphere of the Fixed Stars. Thirdly, in reflection on its own substance as a matter of possible existence, the First Intellect became the cause of the existence of the substance of the Zodiacal sphere itself. The body of this sphere was then created by means of a double cognition of the Universal Soul. Its reflection on the substance of the First Intellect, realising the perfection of this substance, brought into existence the form of the sphere; and, in thinking of its own substance, the Universal Soul realised its imperfection and with this brought the matter of this sphere into being. This process was continued in exactly the same way: the Second Intellect caused the existence of the Third Intellect, the soul and the substance of the Sphere of Saturn (*zuhāl*), whereby its body was again generated by the twofold reflection of the soul of this sphere, and so on. Finally, the Ninth Intellect, belonging to the sphere of the moon, produced the Tenth Intellect, which governs the world of the elements and natural objects. The Ninth Intellect is also referred to as 'Active Intellect', *al-al akl-fa "āl*, and holds a special position in so far as it is the one that acts directly on the world of elements, bringing them from potentiality to actuality.

The following three features are characteristic of this cosmology of intelligences: 1. To each sphere an intellect is attached which brings into existence the intellect, the soul and the substance of the sphere below it. 2. Each sphere thus contains the principle of the one below it. 3. Intellect and Soul are the moving agents of the spheres.

Let us look more closely at this third feature. Although both intellect and soul are moving forces their functions must not be identified. Naṣīr al-Dīn explained their relation with great clarity:

'Each sphere was provided with a soul and an Intellect as the controller of the latter. This was done in order that each sphere should possess an independent and a direct moving agent. This independent moving agent was the Intellect and the direct moving force was the soul. This may be compared, for example, with a magnet, which itself does not move but which brings iron into motion and attracts it to itself. And the 'direct mover' is like, for example, the wind which whirls round a tree and shakes it.'[50]

For the majority of Muslim authors this intellect-soul relation was that of a pair of opposites, whereby the Intellect became the active principle and the soul the passive. In further analogy the Intellect emerged as the masculine factor and the soul as the feminine. Another system of symbols, the roots of which were to be found in Hermetic texts, was mainly employed by Islamic mystics like Ibn-al-'Arabī. The Intellect became identified with the Divine Breath that stirs the soul to motion, which the soul in turn transmits to its proper sphere.[51]

This dualism was further developed by the theory of alchemy. According to this spiritual alchemy the traditional basic substances, sulphur and mercury, were identified with the active and the passive principle, masculine and feminine, spirit and soul, Divine Act and Universal Nature. The motor which operates this system of cosmic polarities, which all ultimately represent the same cosmic principle at different levels of modality, is their sympathetic attitude towards each other. Sympathy between two contrasting terms leads to synthesis which generates another pair of opposites at a different level. This process is not, however, carried on infinitely, but finds its end and culmination in a being, which holds a markedly distinct position within the whole universe: Man.

The concept of Universal Man as a microcosm is not an invention of the Muslim philosophers. It was the current view amongst the Neo-Platonists and appears also in the *Corpus Hermeticum*.[52] Islamic philosophical and scientific thought, with its clearly anthropocentric orientation, was thus well prepared to assimilate and incorporate such kind of doctrine.

Within the ontological chain of being, man occupies the last stage. It has been said already that the stages of existence were arranged according to increasing degrees of multiplicity. The level of greatest multiplicity is at the same time the lowest, that of man. This accounts for man's unique position. In his reality all the cosmic principles are combined in a perfect synthesis. Because of this, his knowledge is superior to the knowledge of every other being. This has led to the concept of Perfect Man (*al-insān al-kāmil*). He reflects the totality of the cosmos, and, since the cosmos is a manifestation of God's attributes, Man in his cosmic function is the most perfect representation of the Divine Reality. Man is to God like a mirror in which He can contemplate His own qualities. With Man, the creation is unified and traced back to its common single source, back from multiplicity to the unity of the Divine Being. When man becomes aware of his twofold nature, of that outer aspect which connects him with the world of material bodies, and the inner aspect, the reflection of the Divine Being, this process of gradual realisation constitutes his spiritual *mi 'rādj*.

'Abdul Karīm Djīlī, a well-known disciple of Ibn al-'Arabī, in his treatise on Perfect Man tells us: 'Know that Universal Man bears within himself correspondences with all the realities of existence. He corresponds to the superior realities by his subtle nature, and to the inferior realities by his gross nature.'[53]

This particular cosmic view, which is expounded in a masterly way by important Muslim philosophers, proves the original character of Muslim scientific speculation, and offers a noteworthy model of the universe.

Notes

1. On the transmission of scientific sources from Greek into Arabic, and hence into Latin, see R. Walzer, 'Greek into Arabic, Essays on Islamic Philosophy', *Oriental Studies* I (Oxford, 1962); F. E. Peters, 'Aristotle and the Arabs; The Aristotelian Tradition in Islam,' *Studies in Near Eastern Civilisation* I (New York, 1968); L. Gardet, 'Le Problème de la 'philosophie musulmane'', *Mélanges offerts à Etienne Gilson* (Paris, 1959).
2. See W. Montgomery Watt, *Islamic Philosophy and Theology*,

Edinburgh 1962, pp. 37–48; M. Meyerhof, 'Von Alexandrien nach Bagdad', *Sitzungsberichte der Preuss. Akad. der Wissenschaften*, Phil.-hist. Klasse (Berlin, 1930), 389–429.

3. Koran 2, 29–33, in Arberry's translation.

4. Ikhwān al-Ṣafā, *Rasā' il*, Beirut 1957, II, 23–6, 456–73. Ibn Sīnā, *Risāla fi 'l-'ishk*, tr. E. L. Fackenheim, *Medieval Studies*, Toronto, VII (1945), 208–28.

'Umar Khayyām, *Raudhat al-kulūb*, tr. A. Christensen ('Un traité de métaphysique de 'Omar Hayyam), Le Monde Oriental, Uppsala, I (1906), 5–13.

Ibn al– 'Arabī, *Futūḥāt al-Makkīya* (Cairo, 1976), II, 373 ff; III, 356 ff.

5. For example, the so-called 'Theology of Aristotle'.

6. C.f. al–Ash 'arī, *Makā lāt al-Islāmiyīn wa Ikhtilāf al-Musallin* H. Ritter (ed) (Istanbul, 1929), I.

7. Koran: 2, 17; 6, 102; 12, 26; 16, 38 and others.

8. Koran: 57, 3.

9. Koran: 11, 7.

10. Koran: 21, 30; 50, 34.

11. Koran: 41, 9–12.

12. Koran: 25, 61–2; 50, 5–6; 40, 26; 17, 12.

13. Koran: 21, 30.

14. Koran: 22, 64.

15. Koran: 35, 65.

16. Koran: 13, 2.

17. Koran: 65, 12; 78, 12.

18. See al–Thaālabī, *Ḳiṣaṣ* 4.

19. Koran: 2, 55.

20. As distinguished from the earthly Paradise, the Garden of Eden.

21. Koran: 53, 14.

22. See H. Corbin: 'Épiphanie divine et naissance spirituelle dans la Gnose ismaèlienne', *Eranos Jahrbuch*, XXIII (1954), pp. 141–250; H. Corbin: 'Les motifs zoroastriens dans la philosophie de Sohrawardī Shaykh-ol-Ishraq', *Publications de la Société d'Iranologie*, No. 3, Teheran; S. H. Nasr, *Three Muslim Sages: Avicenna, Suhrawardi, Ibn Arabi* (Cambridge, Mass.), 1964, 70–4.

23. Koran: 17, 1.

24. Miguel Asín Palacios, *La Escatalogia Musulmana en la Divina Comedia* (Madrid, 1919).

25. Koran: 52, 4.

26. Again, this hell has to be distinguished from the earthly hell mentioned above.

27. See Plate 25. There are also very fine examples in the Bibliothèque Nationale in Paris.

28. Koran: 65, 12.

29. v. *supra*, 8.

30. For Ibn Sīnā cf. H. Corbin, *Avicenna and the Visionary Recitals*, Eng. trans. W. Trask (New York, 1960), 171 ff. Suhrawardī wrote a *Risāla fi 'l-mi 'rādj*. See further H. Corbin and P. Kraus, 'Le bruissement de l'aile de Gabriel' (trans.), *Journal Asiatique*, Vol. 52 (1935), 1–82.
Ibn al–'Arabī, At Futūhāt al–Makkīya, II, 356–75; III, 447–65.
al–Ma 'arrī, *Risālat al-Ghufrān* (Cairo, 1950).

31. The order in which they appear corresponds to their natural positions, fire being located at the highest part of the sublunary sphere, underneath which comes air, then water, and finally earth, with the highest degree of compactness.

32. Al–Birūnī, *Chronology of Ancient Nations*, trans. E. C. Sachau (London, 1879), 248.

33. See S. H. Nasr, *Science and Civilisation in Islam* (Chicago 1964), 346.

34. For al–Ḳazwīnī, see H. Ethé, *Zakharia ben Muhamman El-Ḳaẓwinis Kosmographie* (Leipzig, 1868).

35. Ikhwān al–Ṣafā, *Rasā' il*, II, p. 39, and *Risālat al-Djāmi 'a*, Damascus (1949), I, p. 28; T. Burckhardt, 'Clé spirituelle de l'astrologie musulmane d'après Mohyiddin ibn Arabi,' *Etudes Traditionelles* (Paris, 1950).

36. The theory of illumination was developed, particularly by Suhrawardī, into a philosophical doctrine which became of paramount importance in the Islamic East in the later thirteenth and fourteenth centuries, and was known as *al-ḥikmat al-mashriḳīya*.

37. See A. Mieli, *La science arabe* (Leiden, 1938), 87 ff; S. H. Nasr, *Science and Civilisation*, 170.

38. Al–Taṣawwurāt, trans. W. Ivanov. *The Ismaili Society Series A*, No. 4 (Bombay, 1950), 15.

39. See Chapter 8.

40. See K. Kohl, 'Über den Aufbau der Welt nach Ibn al–Haitham,' *Sitzungsberichte der Physikalisch-mediẓinischen Soẓietaet in Erlangen*, Vol. 54 (1923), 140–79; A. Mieli, *op. cit.*, 87–8.

41. See G. Sarton, *Introduction to the History of Science*, Baltimore 1929, II, 16.

42. The vacuum was employed by Islamic physicists in a purely hypothetical sense, since they did not believe in its physical possibility.

43. Al–Bīrūnī, *Istiʿāb*, trans. S. H. Barani, 'Al–Biruni's Scientific Achievements,' *Indo-Iranica*, V, No. 4, 1952, 276.

44. Al–Bīrūnī: *Istiʿāb*, 277.

45. *Risālat al-Djāmi 'a*, I, 28.

46. *Raudhat al-Ḳulūb*, 12.

47. Al–Fārābī, *Philosophy of Plato and Aristotle*, trans. M. Mahdi (New York, 1962), 19.

48. Not in terms of evaluation but of local position.

49. Died 672 AH/1274 AD.

50. Al-Taṣawwurāt, 13.

51. For example in Ibn al-ʿArabī's treatise, *Nushat al-Ḥakk*; Ms. Istanbul, Shehit Ali Pasha, No. 2813, fol. 4a.

52. A. M. J. Festugière, *La Révélation d'Hermés Trismégiste* (Paris, 1949–54), II, 'Le dieu cosmique'.

53. Translation from Nicholson, *Studies in Islamic Mysticism*, 105.

7

Scandinavian Cosmology

H. R. ELLIS DAVIDSON
Lecturer at Cavendish College, University of Cambridge

The youngest of the cosmologies which we are considering, that of pagan Scandinavia, developed out of a system of religious belief which can be traced back to the Scandinavian Bronze Age, perhaps as far back as 1600 BC, and which continued until the coming of Christianity. The conversion of Northern Europe came late: in the year AD 1000 in Iceland, a little earlier in Norway and Denmark, and considerably later in Sweden; and the last period of the pagan religion was a vigorous one, to judge from its manifestations in literature and art. The cosmology as we have it developed during the Viking Age, which may be dated roughly AD 750–1070. Clearly the Vikings in their western adventures had many encounters with the Christian church, hostile and violent although these may often have been, while in eastern Europe on the other hand they were in close contact with pagan and barbaric peoples over a long period. Their cosmology shows signs of influence from both directions, while it retains its basic Germanic character. It is a simple one in comparison with some of the elaborate schemes of thought in the Ancient World; it arises out of an oral culture, where writing was confined to brief runic inscriptions, and it was never developed by a highly organised priesthood or by schools of trained philosophers. Yet I would claim that it is neither naïve nor merely derivative; on the contrary it emerges as a rich and individual cosmology, both in spirit and symbolism.

To rediscover it, it is necessary to range widely through many different kinds of evidence, including that of literature, philology, archaeology, folklore and religious symbolism in Scandinavian art. The mythological literature was for the most part preserved in Iceland; it includes some mythological poems and many prose stories about the gods and the Other World, in Icelandic and Latin. It was recorded in monasteries, and presents the usual difficulties familiar to all who deal with religious concepts recorded by men of a different culture, although happily the enthusiasm of the Icelanders for their early traditions led them to preserve far more than one might reasonably expect. One of their most gifted thirteenth-century statesmen and writers, Snorri Sturluson, went so far as to compile the *Prose Edda*, a handbook for poets on the world of the pagan gods, giving explanations of metaphors based on the old myths. The first part was an account of cosmology, and he presented this in the form of dialogue between an early Swedish king called Gylfi, with an insatiable thirst for information, and three mysterious figures called High, Just-as-High, and Third. These reply to the King's questions with commendable patience, outlining the story of the divine world from its creation to its final destruction. Much of Snorri's material comes direct from poems, but many of his sources are lost; he is apt to put together scattered pieces of evidence to form a logical whole, and one of the main problems is how far, two hundred years after the conversion, he has deliberately shaped his material. However, there is no doubt that we owe him an immense debt for recording so much for the instruction of King Gylfi and other inquisitive mortals, writing moreover in superb Icelandic prose and with much enthusiasm.

Beside the evidence of literature, place-names, names and titles of the gods and runic inscriptions of the pre-Christian period, we have much archaeological evidence, the result of a long and excellent record of work in Scandinavia on the rich material available. This goes far beyond the period of literary records, extending into the second millennium BC, when we are ignorant of what language the inhabitants of Scandinavia spoke. Throughout the pagan period there were many changes and developments as new influences came in from outside and were absorbed into the religious system, but certain elements had an extremely long life. I think it possible that a man returning from the early Bronze Age would have found much that was still familiar in the religious practices and symbols of the Vikings.

On the other hand, the mythology is not simple; we are presented with a pantheon of gods and goddesses with a great many names and perplexing characteristics. This is the result of long development, and of the inability of those who recorded the myths to sort out the confusion and distinguish between gods worshipped by the people and those who were mere literary figures or doublets of the great deities under different names. Four major deities stand out from the rest: Odin, god of battle, magic, inspiration and the dead; Thor, the god of thunder and ruler of the sky; and Freyr, with his sister Freyja, twin deities of fertility. Although I shall try to avoid confusion in this account, it will be necessary to name others from time to time when they play some part in the cosmology. Here as elsewhere, we must be prepared for the gods to trespass on one another's territories; Thor was a sky god, worshipped in western Scandinavia in particular, whose cult was carried by Norwegians into Iceland; Odin was venerated particularly in eastern Scandinavia, and although he began as a ruler of the land of the dead, he also took over many functions of the god of the sky. There was no one great centre where the mythology could be unified and organised, for Scandinavia was made up of many small, isolated communities, each of which might develop its own particular slant on the religious system. Yet in spite of this confusion, it is possible, or so I trust, to present a fairly consistent picture of Scandinavian cosmology.

It is concerned with the creation of the worlds, with their relationship to one another, and with their eventual destruction and re-creation. The Icelanders were greatly interested in such concepts, and certain poems of the question and answer type imply the existence of an extensive body of knowledge surviving into the tenth century. At its simplest the pre-Christian view of the cosmos might be represented as a mighty tree in the centre of a round disc surrounded by ocean: 'the bright plain which the water encircles', as it is expressed in the Creation Song in the Anglo-Saxon poem *Beowulf*.[1] There is a spring under the tree, and its roots go down deep into the regions below. Many beings dwell in and beneath it, and a great serpent lies in the depths, curled round the circle of the world. The idea of the encircling sea was natural to a people many of whom saw the sun go down beneath the western ocean or rise above the Baltic, whose land was cut into by many fiords and rivers and bordered everywhere by islands, so that they had to rely on boats as their main means of communication and trade. In Norway and much of Sweden the way inland was blocked by huge mountains, while in Iceland the great lava wastes of the centre continually encroached on the coastal lands and fertile valleys. The great central Tree could hardly have evolved in treeless Iceland, however; it could have come from the peoples of the thick forests of the southern Baltic, or the oak woods of Germany where the sky god was worshipped in forest clearings, while it is probably significant that the peoples of north-eastern Europe and Siberia also have a World Tree firmly established in the midst of their religious imagery.

Whatever its origin, the Tree looms large in the Icelandic mythological poems, although it is rare in northern art. The nearest parallel which I can find to the literary picture of the world is on a great memorial stone found under a church floor in Sanda on the island of Gotland in 1954, which dates back probably to the early sixth century AD (see plate 26). Here there is a tree in the centre of the stone, faint but unmistakable, with a dragon or serpent creature beneath it. The great whirling disc would seem to represent the moving circle of the cosmos, including both the conception of the turning year and the revolving circle of the sky across which the sun passes and the constellations are set. Other ways of representing this were the wheel and the swastika. This particular motif probably came from south-eastern Europe, and it is occasionally found on Roman tombstones; it is used on a number of sixth-century stones in Gotland, which were originally painted in bright colours.)[2] Below the large disc on the Sanda stone are two smaller circles, each with an encircling serpent, although on other stones one serpent may be made to envelop the two circles in a figure of eight; these circles probably represent day and night, corresponding to the sun and moon on Roman tombstones.

Below the tree and the serpent is a ship, the long-enduring symbol in Scandinavia of the voyage of the dead. This can be traced back to the Bronze Age, when the journey of the sun across the heavens by day seems to be represented in the North by a wagon drawn by horses, and that through the underworld at night by a ship.[3] By the Viking Age we have only occasional references to the chariots of Sun and Moon, and the wagon symbol has passed to Thor, the Thunder God, who rattles across the heavens in a wagon drawn by goats, its wheels striking out sparks as it goes, a homely but powerful image. The funeral ship however continued to the end of the pagan period as a dominant symbol; a ship or boat was used to hold the bodies both of royal dead and humble folk in Norway, Sweden and eastern England; sometimes the boat formed a funeral pyre on which the dead was burned with elaborate ceremonial, while it might be outlined over a grave in stones, or represented in funeral carvings.[4] On one stone from Gotland we have the funeral ship with a spear passing over it, symbol of the god of the dead who marked it for his own.[5]

In 1927 Hocart, in his book on *Kingship*,[6] commented on the Scandinavian world-picture, that of the tree on the circular disc, as being such a bad one that it could only be accounted for by the conception of a circular burial mound surrounded by a ditch and surmounted by a sacred tree, thus supporting his theory that ideas of the Other World were based on the local form of the tomb. He was, however, wildly over-simplifying an already simplified picture which he had obtained from Grimm's *Teutonic Mythology*, and like many others he was requiring from an ancient cosmology the kind of diagrammatic logic familiar in western thought. The burial mound was certainly of great importance in Scandinavia and Germany from very early times, but we have no evidence that a tree was regularly planted on it. However, it was customary up to the last century to have a guardian or lucky tree planted beside the family dwelling place, on whose well-being that of the family depended. MacCulloch gives an example of a birch tree in western Norway which in fact stood on a mound; this fell in 1874, but while it still lived no one was allowed to cut the branches, and members of the family poured ale over its roots at Christmas and other festivals.[7] Such practices may have helped to influence descriptions of the World Tree in the literature, for sometimes it seems to be represented as the guardian tree of the gods.

Of greater significance, however, is the fact that the Tree marked the place where the gods met to take counsel together and to frame laws, riding to the meeting-place on their swift horses, while Thor, who was no rider, waded through the rivers which lay in his path. In Iceland the place where the annual Law Assembly was held, Thingvellir, was a spot where many roads met, to which men journeyed from all parts of the island. It had a sacred pool and a place of sacrifice, and at the Law Rock, the spot where a great fissure divided the plain, the law was recited to the assembled representatives of the people. Iceland was not colonised until the late ninth century, but a more ancient sanctuary at Gamla Uppsala in Sweden was also a place of assembly, marked by a mound, while beside this stood a row of three more huge mounds which were the burial places of early Swedish kings, and many smaller ones beyond. The shrine of the gods probably stood on the site of the little medieval church which still survives, and there was a sacred spring and a grove of trees on which sacrifices hung at the great festivals.

M

It seems unlikely that there was a special sacred tree at Thingvellir, but there may well have been one at Uppsala, since in a paragraph added to the history of Adam of Bremen, written about 1075, it is stated that a huge tree stood beside the temple of the gods; its branches stretched afar, and it remained green both summer and winter, no one knowing what kind of tree it was. There were also sacred trees in northern Germany, great oaks believed to be under the special protection of the thunder god. Earnest Christian missionaries like St Boniface cut them down, to the terror and rage of the people. The Icelandic poets, however, describe the World Tree not as an oak but an ash, while some scholars argue that it must originally have been a yew, because of the emphasis on perpetual greenery and the importance of the yew in later Germanic and Celtic tradition.

In any case, the Tree where the gods assembled recalls the importance of the tribal centre, the spot on earth where men could come into contact with the gods and which formed the mid-point of their own little world. Whether the symbol of a central pillar or a cosmic mountain was earlier than that of the tree is difficult to establish. The pagan Saxons had a pillar called Irminsul, which in a ninth-century account is described as the column of the universe, upholding all things.[8] The high seat pillars of Icelandic and Norwegian halls were in themselves important symbols, and specially associated with the god Thor; when Norwegians left home to settle in Iceland, they were said sometimes to take these pillars with them and to throw them into the sea as they neared the island, so that Thor could decide where to come to land. These pillars were probably the great tree-trunks which took the main weight of the building,[9] and it is possible that in pagan shrines such as that at Uppsala it was a pillar of this kind and not a living tree which represented the World Tree, forming a link between men and the god of the sky.

The World Tree had many names, but the one most used in the poems is Yggdrasil, and one of Odin's names was *Yggr* The most satisfactory interpretation of Yggdrasil seems to be

Horse of Yggr. The gallows was described as a horse on which the hanged man rode, so that Odin may have been thought to ride on the World Tree in the sense that he was represented hanging from it as a sacrifice, pierced with a spear like one of his own victims, 'Odin given to Odin'.[10] Such a hanging was a grim initiation rite, by which the god obtained knowledge of the future.[11] By the late Viking Age Odin had taken over many functions of the sky god, and his association with the Tree may be one of these and so be relatively late. However, there is little doubt that the links between Thor and the oak and Odin and the ash are based on beliefs which go very far back into the pagan past in both Germany and Scandinavia.

Another way in which Odin obtained supernatural wisdom was by drinking from the spring beneath the Tree, the Well of Mimir; but for this privilege he had to give up one of his eyes as payment. Mimir is represented as an ancient giant, guardian of the sacred well; he was beheaded by the gods, and after his death Odin embalmed his head and kept it so that he might consult with it when he was in urgent need of counsel. Thus the Tree was repeatedly associated with the gaining of wisdom, and it was also thought of as the source of life. It was inhabited and surrounded by living creatures: high on its branches sat an eagle, and on its summit a cock, glittering like gold, which could give warning if danger threatened the gods. The serpent Nidhogg lay beneath its roots, and sometimes we hear of a host of serpents. Harts and goats fed on the branches of the Tree, one splendid hart in particular from whose horns flowed a river which fed all the streams of the world. In one poem we are told that the Tree endured unimaginable anguish, since

> a hart gnaws it on high, it rots at the side
> and Nidhogg devours it below.[12]

Yet it remained for ever green and living, and neither weapons nor fire could destroy it; it embodied continual destruction and creation, fit symbol of the life of man continuing through the generations.

Something of this conception seems to inspire the ornament

like that on panels from an early stave church at Urnes in Norway, showing restless creatures biting at foliage and at one another. The spring beneath the Tree contributed to its life; we may assume that there was only one spring, bearing different names, although Snorri tells us that there was one under each of the three roots. One name was the Well of Urd, and Urd was the foremost of three maidens who determined the fates of gods and men and are called the Norns; they are said to sprinkle the Tree with water from the well. In one obscure passage the fruits of the Tree are said to help women in childbirth,[13] and this recalls the idea among Siberian peoples that the World Tree was the home of unborn souls. Dew fell from its leaves to nourish the world beneath, and at the destruction of the world two living creatures will be kept alive within it and fed on its dews, so that they can afterwards go out and repeople the earth; one of the Tree's many names was Shelterer. From the udders of the goat which gnawed its branches mead flowed instead of milk, providing drink for those in Odin's hall. Two swans were nourished in the spring, and this image seems likely to be a very ancient one going back to the Bronze Age; while between the serpent under the roots and the eagle in the branches passed a nimble squirrel bearing messages, which I take to be hostile ones, from one to the other. The contest between the eagle, bird of heaven, and the serpent, symbol of the under-world, is another ancient and widespread image, and a battle between these two creatures is a possibility in the natural world, filmed most impressively by Disney in *The Living Desert*.

We are told in the poems that the Tree had three mighty roots to support it, and that these stretched into three different realms, those of Midgard, the Middle Stronghold, home of mankind, Jotunheim, realm of the Frost-giants, and Hel, the land of the dead. The implication is that these lie side by side and that the stronghold of the gods, Asgard, is above them. Snorri, who like us was often perplexed by what he found in the poems, placed the land of the gods under one of the three roots of the Tree instead of that of man (which presumably he placed in the centre), but this brought new difficulties, as he had to imagine their root of the Tree in the sky. No doubt different conceptions existed at different times; certainly Asgard, stronghold of the gods, was sometimes thought of as being above the earth, for a bridge called Bifrost, which Snorri calls the rainbow bridge but which is also said to be a bridge of fire, stretched from earth to heaven. This the gods used on their journeyings, but it was doomed to be shattered by the enemy host which was to cross it on the last day. Because of the danger of attack, a guardian was stationed beside the bridge, the god Heimdall, who bore a great horn to give warning to the gods. He was an untiring sentinel who needed less sleep than a bird, whose ears were so keen that he could hear grass growing on earth or wool on a sheep's back. It was evidently from below that danger threatened Asgard, and Odin's seat, Hlidskjalf, seems to have been up in the Tree itself, since from it he could look down into all worlds.

It is possible that beside this picture of worlds stretching round and beneath the Tree there was at one time a conception of a vertical structure of several worlds extending one above the other. In one of the mythological poems[14] a seeress declared that she remembered nine worlds 'in the Tree', but like much else in the poems this is hard to interpret, and the meaning may be that the nine worlds lay among the Tree's roots, or even that they were worlds succeeding one another in time. There are cases also where supernatural beings claim that they have travelled through nine worlds, but it is not clear whether this is a horizontal or a vertical journey. Snorri tells us that Hel, the ruler of the land of the dead, had authority over nine worlds, and of these one at least, Niflheim, is said to be below Hel. The Viking world was familiar with the symbol of a mound which rose in steps, as at Tynwald Hill in the Isle of Man, the place where the Manx parliament still assembles and which goes back to the time of the Viking kingdom there (see plate 31). The hill has four levels, and the representative of the reigning monarch now sits on the summit (earlier this was the seat of the Lord of the Isles) together with the Bishop, while the two Deemsters sit below, and below them again the twenty-four members of the House of Keys.

There is an early tale of ninth-century Norway, in which a minor king diplomatically resigned to a more powerful one by rolling himself down a mound from the place of the kings to that of the jarls below.[15] In these cases the emphasis is on law and on the structure of society, but law also, it must be remembered, was under the will of the gods. If indeed the tall narrow stave churches of early Christian Norway, with their series of roofs (see plate 29), were based on earlier wooden shrines of the pagan gods, then it is possible that this pattern represents a vertical picture of the cosmos. It is also interesting to note that in a picture in an early manuscript of Snorri's *Prose Edda* the three powers who were consulted by King Gylfi were raised one above the other (see plate 28).[16]

In any case it is clear that we must abandon all attempts to form a clear, neat diagram of the Scandinavian cosmos as Hocart wished to do. This is not the impression left by a study of the literature. The myths have a sense of mighty spaces, of darkness and terror and tremendous barriers confronting travellers, human or divine, on journeys from one realm to another. There are great mountains and rushing rivers, like those of Norway and Iceland, blocking the road. The feeling of immensity, darkness and cold is in keeping with the long cruel Scandinavian winters with their brief hours of daylight and savage blizzards. Yet as if reflecting the restlessness of the Viking Age, there is constant journeying between the worlds. The road from the land of the gods to that of the dead took nine nights through deep, dark valleys, over rivers welling up from the lowest depths, across a bridge which resounded under the marching feet of the dead, and finally downwards and northwards to the mighty gate of Hel, Helgrind.

The journey to Giantland was similarly long and dangerous, and seems to lead eastwards; Thor made regular expeditions there to kill trolls, wading through rivers and passing through mighty forests on the way. Odin, attended by his ravens, rode far over land and sea to visit the battlefields of earth; he entered the halls of kings disguised as an old man in a broad-brimmed hat and cloak, or ventured into the land of the giants for some cunning purpose of his own, or passed down into the land of the dead past the dog guarding the entrance. He or his messengers escorted kings and heroes who fell on earth to his own realm of Valhall, the Hall of the Slain; one scene on a helmet plate shows one of these emissaries guiding the spear of the warrior whose horse is being stabbed from below, and who is therefore presumably about to depart for Valhall.[17]

We hear also of journeys of young gods or heroes through realms of darkness and a wall of fire to win a bride from the underworld. There are strange tales of mortals travelling to the realm of a fair giant called Gudmund, lord of the Shining Plains, the land of the Not-Dead, while his neighbour, Geirrod the giant, rules a dark and sinister land of death; such stories have resemblances both to Celtic tales of the other world, and also to shamanistic lore among Finno-Ugric and Siberian peoples; there is, moreover, much in the stories and poems to emphasise the journey of the disembodied spirit, in bird or animal form, through far distant realms while the body lies in a trance.

While the four kingdoms of gods, giants, the dead and mankind remain consistent, other realms are mentioned in the literature. In one poem, *Alvíssmál*, the names for the forces of nature used by the inhabitants of different realms are listed. For instance, in reply to the question of what names are given to the wind among the various beings, the reply is:

> It is Wind among men, Waverer to the gods [a name for Odin]
> Neighing One to the mighty powers;
> Shrieker to the giants, Whistler to the elves,
> and in Hel they say Breath of Storm.

The names of the beings in different verses vary slightly, and it is possible to build up a tentative list of nine different kinds: one, the Mighty Powers; two and three, the two races of the Gods, the Aesir and the Vanir; four, the Giants; five, the Elves; six, the Dwarfs; seven, the Dead in Hel; eight, the Heroes dwelling with Odin; and nine, Mankind. The Mighty Powers, the name for which is *Regin*, only used in the plural, would seem to be gods or powers who are makers or rulers; Ragnarok, the final catastrophe, means literally Doom of the Powers. There is a general implication that beyond Odin and Thor there is a greater power, Necessity or Fate, or as the Anglo-Saxons called it, Wyrd: 'Wyrd is stronger and God mightier than any man may imagine', as it is expressed in the Christian poem *The Sea-Farer*. To trace back the conception of fate in Scandinavian pagan thought as the power above the gods Odin and Thor, who themselves were doomed to destruction, is not precisely an easy task, nor is it easy to decide how ancient is the idea of the Norns as the beings who decide what is to come; or whether there was an early conception of one supreme creator god or gods above the rest. It is generally recognised, however, that the idea of fate is a powerful one in Scandinavian literature.

The two groups of gods known as the Aesir and the Vanir are familiar from the numerous myths in which they play a part. The Aesir is the group associated with the sky, and the outstanding members of this group are Odin, Thor and Tyr. Each has his own dwelling, but they also form a pantheon, living in the stronghold of Asgard. The Vanir are often represented along with the Aesir, but they have a realm of their own, Vanaheim, and a myth survives of a war between the two groups in the beginning, after which a truce was made and hostages exchanged. The most famous of the Vanir are the twin fertility deities, Freyr and Freyja, and their father Njord, who in Viking times was seen as a god associated with ships and the sea. There are many other gods and goddesses whose names are given, but this may be due simply to many communities each possessing its own local fertility pair, so that Freyr and Freyja, whose names simply mean Lord and Lady, reappear under many different titles. Njord may originally have been the male partner of the goddess Nerthus, Mother Earth, whose worship in Denmark in the first century AD is described in the *Germania* of Tacitus, Chapter 40. There is some indication that the dwelling of the Vanir deities was not in the sky but in the depths of earth and sea, and the fair giantesses of the underworld, who mated with the sky gods, could be manifestations of the goddess Freyja, who, as Snorri wisely said, was the last of all the deities to perish. The division between the two main groups may be traced back to the Bronze Age, where there is some evidence for the conception of a god of the sky and a goddess of the earth.

The Vanir, however, must be firmly differentiated from the Frost-giants, who were utterly hostile to the gods, although they longed to carry off the goddesses for themselves. They struggled continually to defeat the attempts of gods and men to create a bright, settled abode in the midst of darkness and chaos. If they had their way, then Asgard would be overthrown, the sun and moon destroyed, and men's homes laid waste, and in this they would be aided by the monsters whom the gods kept in check. The wolf Fenrir was bound by the god Tyr, who seems to have been an early Germanic sky god, and probably the predecessor of Odin as god of battle. As Odin gave an eye to obtain wisdom, so Tyr sacrificed a hand to bind the monster who would have destroyed the gods. The World Serpent, too, was secured in the depths of the sea, for if he broke loose and came on land, the sea would overwhelm the earth. One of the most famous myths, celebrated in a number of poems of the Viking Age and carved on monuments of Scandinavia and Viking England, was that of Thor fishing up the serpent to strike it down with his hammer. It became one of the many exploits of the red-bearded god, humorously and vigorously told, but it seems probable that it was a very ancient myth, representing the primeval struggle of the sky god with a monster of chaos, and that when Thor struck the serpent back into the deep this was a memory of the casting of the serpent down into the depths at the beginning, mentioned in the accounts of creation.

The great weapons of the gods, the spear of Odin and the hammer of Thor, can be traced back into much earlier times, though those who then wielded them no doubt bore different names. Once the giants stole the hammer of Thor, and there is a splendid comic poem *Thrymskvida*, telling of its recovery. In the tenth century the hammer was worn as an amulet, a kind of counter-symbol to the Christian cross, and it seems to have been a powerful symbol in pagan mythology, the weapon of the creator god against chaos; gods and men relied on its protection.[18] In the early period, however, the threat of the devouring monster goes unrepresented in art, and we do not know enough about the religion of the Bronze Age to understand why this should be so. Monsters first appear in Scandinavian art about AD 500, possibly as part of the cult of Odin, which was gaining power about this time.

As the god of the dead, inspiration and secret wisdom, Odin came into rivalry with the powers on the earth, the giants and dwarfs. Sometimes he or Thor competed with them in tests of esoteric knowledge, the gods triumphing in the end by keeping their opponents so engrossed in the game of question and answer that they were caught by the rays of the rising sun and turned to stone. Only by ruthless cunning did Odin succeed in getting the mead of inspiration out of the possession of dwarfs and giants in the mountains and bringing it back to Asgard. The mysterious elves, to whom sacrifices were made and who were said to dwell in mounds, have some links with the Vanir, and possess also some of the characteristics of the dead within the earth. They belong to the more popular world of the supernatural, along with other spirits of the brownie type, dwelling in mounds and stones and waterfalls, and to the landspirits said to help men in farming, hunting and fishing, able to bring bad luck on those who offended them. This popular world of helpful and malevolent spirits lies behind the more complex and lofty mythology of the poems and carvings, and such beliefs probably continued from the earliest times up to the conversion and even after it. Dwarfs, elves, giants and the Vanir all have a footing in it, while they also pass in and out of the more dignified myths of northern paganism.

The dead are said to dwell in the hall of Hel, goddess of death; this sometimes seems no more than a metaphorical statement, but on the other hand there was an ancient conception of an Earth Mother who received the dead, and both Ran, goddess of the sea who welcomed drowned sailors, and Freyja herself sometimes appear in this capacity. There are many contradictory ideas about the dead; they may be said to dwell in their burial mounds as in a house, or walk forth from them to harm the living; some families in Iceland were said to 'die into the mountains' and join their ancestors there; there is the conception also of a green land of the gods where the ancestors feast together. There is some idea of a punishment for the wicked, especially for such crimes as oath-breaking and treachery, while somewhere there is a golden hall for righteous men. These ideas may be due to Christian influences, but not necessarily so. There is also mention of a second death out of Hel into Niflheim, realm of mist, into what seems complete negation; from this deep underworld lying beneath Hel, flow the rivers which divide up the realm of the dead. There are references also to rebirth of the dead into the world of men, emphasised by the custom of naming a child after a dead kinsman, which was carefully observed.

While such ideas about the dead are not worked up into any kind of systematic teaching, the belief in Valhall, the Hall of the Slain, is a more consistent one, although it seems to have been restricted to a small minority, aristocratic warriors who followed the cult of Odin. It is well known because such men were the articulate members of the community, and the poets and artists who worked for them put such ideas to enthusiastic use. Those who allowed themselves to be sacrificed to Odin, perhaps by taking their own lives to avoid disgrace, or who died courageously in battle, might be carried by his eight-legged horse, Sleipnir, to Valhall, or summoned thither by his messengers. His hall had hundreds of doors, so that men could flock into it from all parts of the world, and could also pour out to meet sudden attack; it was guarded by wolf and eagle, creatures of the battlefield and symbols of Odin, and lit by gleaming swords. Here the dead were welcomed by valkyries bearing horns of mead, and they entered to a life of heroes, fighting by day and feasting by night on unending supplies of pork and mead which never gave out. Figures of riders and valkyries were carved on memorial stones of the dead in Gotland, and also worn as amulets and placed in Viking Age graves in Sweden (see plate 27).

To some extent Valhall may be seen as the poet's glorification of the warrior's grave, the goal of men who spent their lives in fighting and violence, and it seems to have its origin in ideas of an underworld realm of the dead rather than the bright world of the sky. That it was not entirely the creation of poets, however, may be seen from accounts by Arab and Byzantine writers of the behaviour of Swedish vikings in eastern Europe.[19] They were amazed by the wild ferocity of these men, their anxiety to win glory and die a death which would long be remembered, and their readiness to fall by their own hands rather than be captured. They accounted for such wild reckless-ness by the beliefs these men held in the after-life,[19] and this explanation seems justifiable. Women also are said in Icelandic literature to have been ready to die with their husbands or masters in pursuit of this kind of renown and immortality, and there is some archaeological evidence to support literary tradi-tions for the practice of suttee in the Viking Age. It is possible that contacts with barbaric tribes in the east helped to encour-age such traditions, and certainly it is in Gotland in the eastern Baltic that we find the rich series of carvings showing the departure of dead warriors to the halls of Odin, in the very spirit of heroic literature.

The beginning and end of the worlds of gods and men is a subject which in Scandinavia as elsewhere invited speculation. The emergence of an ordered world out of chaos is continually emphasised in the literature. The original state of formlessness before creation is not normally represented by water – although there is a concept of the earth rising out of the sea – but by a great abyss, Ginnungagap, which appeared empty but which was in reality pregnant with potential life. The old interpreta-tion of this name was Yawning Gap, but a more satisfactory one links it with the idea of deceit through enchantment, which prevents the beholding of things as they are: the abyss seemed empty, but was not really so.[20] How life began is told in what originally may have been a series of independent myths. First layers of ice formed in the void, while from the region to the south sparks and embers arose, and cold and heat worked together to form a living shape, that of a great giant Ymir. His name is thought to mean 'hermaphrodite', 'two-fold being',[21] for he was both male and female; he gave birth to men and giants, for a man and woman came forth from beneath his arms, and a giant was engendered by his two feet.

Ymir was nourished on the milk of a primeval cow, which also performed the work of creation by licking the salty ice-blocks until from them emerged the figures of the sons of Bor; we do not know who or what Bor was, but the three brothers who came out of the ice slew the primeval giant and formed the earth from his body, the sea from his blood, and the sky from his skull. As an additional story of the creation of man, we are told that three creator gods walked on the seashore and came across two trees; in an Icelandic setting it is likely that these were driftwood brought by the sea. Into these they breathed life, giving them form, spirit and understanding, and they became the first man and woman upon earth. The coming together of fire and ice certainly had a special meaning for Icelanders, in a land where few years go by without some volcanic activity which brings blazing lava into contact with snow and glaciers. This unhappily brings death to the land rather than new life, but the birth of the island Surtsey reminds us that a new world can arise out of the sea during an eruption.

It has recently been claimed by German scholars[22] that there is evidence for the concept of a mighty creative deity expressed in symbolic patterns on ornaments of the Germanic pagan period, particularly on the great gilded bronze brooches worn by heathen Anglo-Saxons. Detailed work on a very large number of these brooches from England and the continent show that a favourite motif is a face with an open mouth, from which issues a kind of cloud, which may enclose a disintegrated animal form, such as is characteristic of Anglo-Saxon art of the fifth and early sixth centuries. A Finnish archaeologist who was

interested in this motif compared it with representations of the
Holy Spirit in illuminated Christian manuscripts, where this
is shown as breath or fire;[23] however, the treatment on the
brooches, their date and pagan background, make it very
unlikely here to be Christian in origin, and it may have arisen
out of some fundamental concept in Germanic thought in pre-
Christian times. There are signs of a similar concept on Scandinavian bracteates of the fifth century, little gold amulets imitated from Roman medallions but developed as mythological
designs in northern style. It will be interesting to see further
developments of this theory, the evidence for which is impressive because it is not based on isolated examples but on large
numbers of motifs.

There have been arguments also for the conception of a dual
creation by twin gods in Scandinavian mythology, based
mainly on one very obscure and puzzling poem about two of
the lesser gods, Heimdall and Loki, battling together in the
sea.[24] These two are baffling figures; there is no evidence for
cults connected with them, but they play a large part in the
myths. Heindall, the watchman with his horn, possesses some
of the characteristics of a creator god, and is called the father of
created beings; Loki is a mischievous companion of both Odin
and Thor, but sometimes appears as a sinister giant. He is partly
a friend and partly an enemy of the gods, continually causing
trouble in Asgard, and finally bringing about the death of
Odin's son Balder, after which he was punished by being
bound beneath the earth. He could take on both male and
female form, and in a sense is a creator figure, since he gives
birth to Odin's eight-legged horse, and also begot the Wolf,
the Serpent and Hel, goddess of death; he might be regarded
as a kind of shadow to Odin. The battle in the sea has been
compared with the creation story found in certain areas of the
world, including the Balkans, where two brothers, or in Christian versions God and the Devil, bring up earth to create the
world out of the sea.[25] However, if Heimdall were once a
supreme creator god with Loki as his partner, this had been
almost completely forgotten by the time of the Viking Age.

When Ymir's skull was raised above the earth to make the sky, it was held aloft by four dwarfs at the four points of the compass. It is shown like this on a stone from Cumberland over a Viking grave, the little figures pushing up the sky while at the side monsters, it would seem, try to bring it down.[26] Midgard, the world of men, was protected from the giants by

12. A hog-back tomb, Heysham

a wall made of Ymir's eyebrows. The gods in Asgard persuaded a giant craftsman to build a wall for them too, managing to cheat him of his payment and kill him when the work was done. They sent Night and Day and the fair maiden Sun, with her brother Moon, driving across the sky, and then busied themselves creating laws and making things of beauty and usefulness. The dwarfs, we are told, had bred like maggots in the earth, and some of their craftsmen forged the treasures of the gods. There was the great gold ring, Draupnir, from which nine more rings dropped every nine nights, for Odin, and his great spear, Gungnir; Thor's hammer Mjollnir; the golden boar Gullinborsti and the ship Skidbladnir for Freyr and Freyja – ancient symbols of the fertility deities. Yet there was never real security in Asgard, and as Odin is represented as saying in *Eiríksmál*, a tenth century poem on the death of a Viking king, 'The grey wolf is watching the abodes of the gods'.

It is Odin in particular who plays the central role in the account of the world's destruction. He knew by his foreknowledge that it must come, and so filled his hall with champions so that there would be an ample force ready to meet attack. The sign of the approaching end was the death of his son Balder, sometimes represented as a god and sometimes as an earthly warrior; when he was killed by his brother with a spear, through the cunning of Loki, every attempt by Odin to save him from going down to Hel proved unavailing. Then a terrible winter came upon earth, lasting three years on end with no summer between. Ties of kindred and the rule of law broke down, and at last the sound of Heimdall's horn and the crowing of the cock heralded the approach of the enemy.

There are two aspects of this attack: the marching of an enemy host on Asgard, and the breaking loose of the bound monsters; and scenes from the catastrophe are shown on a number of carved stones of the tenth century in northern England and the Isle of Man, in particular on the Gosforth Cross in Cumberland. It appears that these mythological scenes are intended to symbolise the triumph of Christ over the powers of evil, and thus have been included on memorial stones of the early Christian period both on Man and in northern England. When the signal came, Odin led out his army to the plain where the last battle was to be fought. The enemy were the giants, who crossed the sea in a boat made of dead men's nails, with Loki as their steersman, and the mysterious host of the sons of Muspell, who came from Myrkwood and the region of heat and crossed the bridge Bifrost, breaking it with their weight. The sun and moon had been swallowed by pursuing wolves, and the great monster Fenris-wolf now advanced, his jaws filling all the space between earth and heaven, while the World Serpent emerged from the deep. The Wolf devoured Odin (see plate 30A), though the god was avenged by his young son Vidar, who tore the monster's jaws apart; Thor advanced against the Serpent and struck it down, but was himself slain by its poison. Tyr, a kind of doublet of Odin, killed the hound Garm which guarded the underworld, but was himself killed also. Heimdall and Loki slew one another. Finally the fire giant Surt flung blazing embers over earth and heaven; the sky fell and the stars vanished, fire and smoke rose high, and the world was engulfed by the rising sea.

13. The Gosforth Cross

(A)   (B)   (C)   (D)

But this was not the end. We are told that a new earth arises again from the sea, green and fair. The World Tree has survived untouched, and the sons of the gods who would seem to have been sheltered there along with a man and woman nourished within the Tree during the terrible winter, now come out to begin a new cycle of existence. Once more an eagle is seen in the sky, and a new sun, fairer than her mother, moves across the heavens. As they rest on the fresh grass of the cleansed world, the young gods find the golden playing-pieces which their fathers once used in Asgard.

Here the concept of repeated cycles of existence is clearly stated. How far this may have come in from the East comparatively late in the Viking Age, or how far it was already present in earlier Scandinavian and Germanic religion, is difficult to determine. This is the case also with other concepts connected with the world's end, such as the last battle, the binding of Loki and so on, some of which may have been influenced by Christian legends of Doomsday. While something may be due to literary borrowing, the possibility of oral tradition and the bringing back of legends from eastern Europe must also be taken into account. Certainly the scenes carved on memorial stones make it clear that such ideas were widely held, and were not restricted to the speculations of scholars. The description of the end of the world as a catastrophe in which cold and heat and a tidal wave were all included bears a striking resemblance to what occurred in great volcanic eruptions in Iceland, like that of Hekla in the eighteenth century, and experience of earlier times of terror may have lent vigour to the theme. There is evidence, too, in Danish folklore for widespread beliefs in the eventual flooding of the world by the sea. Here, as elsewhere in the cosmology, we have a foundation of men's instinctive beliefs founded on the observation of their world, on which has been raised the more elaborate superstructure of the myths with their tales of gods and giants.

We know very little of how far a pagan priesthood at centres like Uppsala may have worked on the myths in earlier times, but certainly part of their development was due to poets of the

Viking Age, who delghted to make use of them and create out of them a complex poetic diction. It was due to them, for instance, that the valkyries, who began as fierce elemental spirits of battle, devouring the slain, became dignified ladies in armour, riding on splendid steeds to escort heroes to Odin's hall. More barbaric conceptions seem to have been forced into the opposition, so that demonic troll-women on their wolf steeds and skin-clad giantesses are contrasted with the fair goddesses, who may well have been conceived as such wild figures in earlier times, and who certainly possessed terrible aspects as well as radiant ones when they were concepts of living belief. Funeral ritual and the poetic tributes offered to dead leaders, where they were pictured entering the Other World and being welcomed by the gods, have helped to build up the picture, while there is little doubt that the Scandinavian gift for story-telling and capacity to absorb themes and plots when they came upon them in their travels has helped to enrich cosmology and myth. It has never been possible in this account to get away for very long from the god Odin, and his cult certainly helped to inspire much of the poetry and art which survives. It may be remembered that Odin was the patron of princes, and that both poetry and ancient lore were held to come under his sway.

Although Scandinavian material is late in comparison with other ancient cosmologies, it contains much that is comparable with those of far earlier peoples whose society was based on agriculture, sea-faring and fighting, who lived in small communities under independent leaders who took pride in their ancestry, and who possessed no large towns and no written records. Scandinavia remained outside the influence of classical culture until the coming of Christianity, and suffered no major defeat or national upheaval throughout the pagan period such as would bring destruction to the native traditions. The geographical position of her people and their tendency to travel and explore meant that influences from outside came in continually throughout the pagan period. These were for the most part barbaric ones, since more sophisticated schemes of thought only reached the North by oral tradition, across the intervening area of north-eastern Europe for the most part, by way of tribes and peoples as uncivilised, in the classical sense, as themselves. Symbols and cults reaching the North were absorbed and adapted to the conditions of their own lands and way of life, and brought valuable stimulus to religious thought, and new inspiration to poets, artists and story tellers, without revolutionising what had gone before. The richness of the mythology is largely due to the fact that the Scandinavians were constantly looking outwards, because of their dependence on the sea. While their cosmology has much of the esoteric in its continual invention of names and terms, its runic lore and complex magical ritual, which largely defeats us by its complexity, it was at the same time a system highly relevant to the everyday life and aspirations of men of the Viking Age, a symbolic pattern which served those in the small, hardy communities of Scandinavia well.

Notes

1. *Beowulf* 93.
2. S. Lindqvist, *Gotlands Bildsteine* (2 vols) (Uppsala 1941–2).
3. E. Sprockhoff, 'Nordische Bronzezeit und frühes Griechentum', *Jahrbuch des römisch-germanischen Zentralmuseums*, I (1954), 28–110.
4. H. R. Ellis Davidson, *Pagan Scandinavia* (London, 1967), 113 ff.
5. Found in Stenkyrka Church, Gotland (P. Gelling and H. R. Ellis Davidson, *The Chariot of the Sun* (London, 1969), 148 and Plate 5.
6. A. M. Hocart, *Kingship* (Oxford, 1927), 188.
7. J. A. MacCulloch, *Mythology of all Races*, Vol. II (Eddic), plate XL, 1930. The tree was near a farm at Slinde, Sogn.
8. J. de Vries, 'La Valeur religieuse du mot germanique *irmin*', *Cahiers du Sud*, 314 (1952), 18 ff.
9. H. Liden, 'From Pagan Sanctuary to Christian Church', *Norwegian Archaeological Review* 2 (1969), 18 ff.
10. *Havamal* 138.
11. H. R. Ellis Davidson, *The Battle God of the Vikings*, G. M. Garmonsway Memorial Lecture (York, 1972).

12. *Grímnismál* 36.
13. *Svipdagsmál (Fjölsvinnsmál)* 22.
14. *Völuspá* 2.
15. *Heimskringla, Haralds Saga ins Hárfagra* 8.
16. In a manuscript of the *Prose Edda* in the University Library at Uppsala, reproduced H. R. Ellis Davidson, *Scandinavian Mythology* (London 1969), 15.
17. This scene is found on a brooch from Pliezhausen, on a panel from the Anglo-Saxon helmet from Sutton Hoo, and on another from the helmet in Grave 8, Valsgärde, Sweden. R. L. S. Bruce-Mitford, *The Sutton Hoo Ship-Burial* (Proc. Suffolk Institute of Archaeology 25 (1949) 47 ff; H. R. Ellis Davidson, *Pagan Scandinavia* (London, 1967), 98–9.
18. H. R. Ellis Davidson, 'Thor's Hammer', *Folklore* 76 (1965), 1–15.
19. For example, *Leo the Deacon*, Book VIII: Ibn Miskawaih, *The Eclipse of the Abbasid Caliphate*, trans. Amedroz and Margoliouth (Oxford 1921), 73 ff; for details see H. R. Ellis Davidson, *The Battle God of the Vikings* (note 11 above), 23 ff.
20. J. de Vries, 'Ginnungagap', *Acta Philologica* 5 (1930–4), 41 ff.
21. J. de Vries, *Altnordisches Etymologisches Wörterbuch* (Leiden, 1961), under *twistra*.
22. K. Hauck, *Goldbrakteaten aus Sievern* (Munich, 1970); H. Vierck, 'Ein Relieffibelpaar aus Nordendorf', *Bayerische Vorgeschichtsblätter* 32 (1967), 104–40, and further work included in the above, sections VIII, IX.
23. A. Erä-Esko, *Germanic Animal Art of Salin's Style I in Finland* (Helsinki, 1965), 101 ff.
24. Put forward by Professor Kurt Schier of Munich in a Colloquium on Mythology held at Clare Hall in Cambridge in 1971, as part of a forthcoming book.
25. M. Dragomanov, *Notes on the Slavic Religio-Ethical Legends, the Dualistic Creation of the World* (Bloomington, Indiana, 1963).
26. Tombstone in the church at Heysham, Lancashire: H. R. Ellis Davidson, *Scandinavian Mythology* (London, 1969), 122–3.

8

Greek Cosmologies

G. E. R. LLOYD

Reader in Ancient Philosophy and Science, University of Cambridge

The field of Greek cosmological thought, from Mycenae to
Byzantium, is immense. I can only select some features of it
that I believe to be particularly interesting or important, but
my selection is partial, certainly in one, and probably in two
senses of that term.[1]

I shall concentrate, to begin with, on three interrelated
themes, the transition from mythology to philosophical cosmology, the pluralism of Greek cosmologies, and the development of critical methods. The transition from mythology to
philosophical cosmology is, of course, a well-worn and much-
abused topic. Studies still appear on the theme, and sometimes
with the title, 'from myth to reason', many of them arguing,
or at least assuming, that the two are directly comparable, and
that the one simply supplanted the other. One associates with
the 'Greek miracle' view of ancient history the tacit assumption
– and sometimes it is not just tacit – of a contrast in kind
between the Greeks and their near eastern neighbours. To begin
with (so this view has it) there were those charming, but child-
ish, Egyptians and Sumerians with their weird and fantastic
notions about the cow-goddess in the sky, the sweet waters
under the earth, and so on, and then along came the Greeks
who were adult rational people like ourselves. The notion that
there is or was a mythological or pre-rational or pre-logical
mentality, different in kind from a scientific, rational or logical
mentality is at best grossly oversimplified and at worst a piece

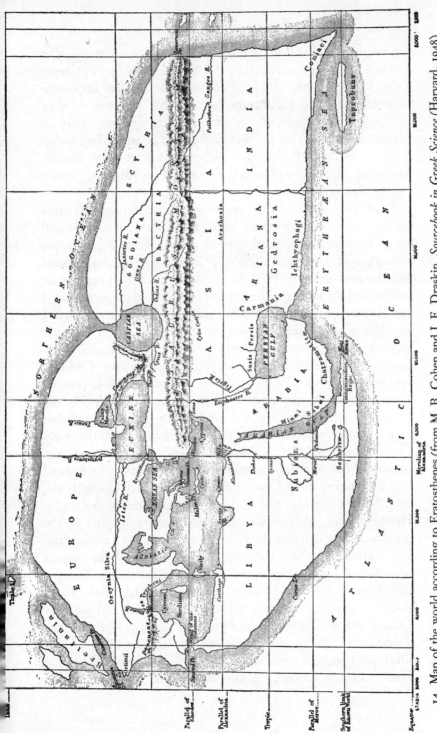

14. Map of the world according to Eratosthenes (from M. R. Cohen and I. E. Draskin, *Sourcebook in Greek Science* (Harvard, 1948)

of dangerously misleading propaganda.² When we study what actually happened in the sixth and fifth centuries BC in Greece (and I am not denying, of course, that certain important changes did take place), the picture that emerges is very different. Indeed one of the main reasons why the study of Greek thought of that period continues to be rewarding is that it provides us with an opportunity to discover what actually happened, and check some of the sometimes wildly exaggerated theses that have been put forward.

I am speaking as if one *can* reconstruct what actually happened, but of course there is a good deal we do not know and many questions we cannot hope to answer. Even the principal doctrines of the three earliest philosophers, Thales, Anaximander and Anaximenes (who worked in the sixth century BC in Miletus) are a matter of controversy. Nevertheless one can, I believe, be reasonably confident about some points. First I would maintain a negative proposition, that in Greece there is nothing we can describe as cosmology in the strictest and fullest sense before the philosophers: by the strictest sense I mean a comprehensive view of the cosmos as an ordered whole (*kosmos* being, indeed, the Greek for order or ordered whole). There are, to be sure, plenty of ideas mentioned by Homer and other sources that are relevant to the question of what the early Greeks believed about themselves, their environment, the origins of certain things and so on. Homer speaks of a river of Ocean running round the world and says that Ocean is the begetter of the gods or of all things.³ Much more interestingly, Hesiod's *Theogony* provides an account of the origins of 'the gods and the earth and the rivers and the immense sea . . . and the shining stars and the wide heaven above'.⁴ This begins with Chaos, in the Greek sense of gap; then Earth, Tartaros and Eros; Erebos and Night, Aither and Day; then Earth produces the Sky (Ouranos), the tall mountains and the sea.⁵ Once Earth and Sky mate and produce Kronos, Rhea and the Titans, we join up with the usual genealogies of the major Olympian gods: Kronos and Rhea are the parents of Zeus and Hera. Outside Hesiod, again, other origin myths involving

Time (Chronos), Night, an original egg, and so on, are known from the evidence for the so-called Orphic and related cosmogonies.[6] These stories include not only theogonical, but also more generally cosmogonical, ideas. One may contrast the ordinary births of the anthropomorphic gods from anthropomorphic parents on the one hand, with, on the other, less straightforwardly anthropomorphic generation, where the figures involved, while undoubtedly divine, are less obviously conceived in human form, or not conceived in human form at all, and where their coming-to-be may or may not be conceived as a birth. When, in Hesiod, Earth bears the Mountains or Dawn brings forth the stars,[7] the words used are still those of human generation ($\gamma\acute{\iota}\epsilon\nu\alpha\tau o$, $\tau\acute{\iota}\kappa\tau\epsilon\nu$) though what is born (the mountains and the stars) may not be conceived anthropomorphically. But, as is well known, there are other early texts relating to an original splitting of Earth and Sky, where there is no question of intercourse or birth, but of a separation of two originally united beings.[8] Yet my main point is this: while these pre-philosophical myths contain a variety of stories of origins, none of them constitutes a cosmology in the strict sense I have defined. None of them, that is, presents a comprehensive account of the world as we know it as an ordered system. Even to say that everything else comes from an egg (or Time or Chaos) does not tell us how the world as we know it is regulated. Cosmology, in the sense of theories concerning how the world as we know it forms an ordered whole, manifesting physical regularities, is a product of Greek philosophy.

One important strand in early Greek cosmological speculation relates to the question of the material constitution of the world. Attempts to give an account of the material constituents of the world may go back to Thales himself. That at least was Aristotle's view, although I am one of those who have reservations about fully accepting his interpretation on this point. The controversy turns on what questions one believes Thales had formulated, and my own view would be that while he may well have considered the problem of what things came from (on which, in an unsystematic way, the myths had already pronounced), there is some doubt about whether he also asked the question of what things as we know them are made of. Yet this dispute is of minor importance for our purposes, since by the time we come to the third of the Milesian philosophers, Anaximenes, we can be fairly confident that he believed that all things are made of the same substance, namely air, which changes into other things by processes which he conceived as condensation and rarefaction.[9]

Early on in Greek philosophy, then, we have evidence of the world conceived as a unity in this sense, that all material objects are varieties of a single substance. But doctrines of how the plural world is an ordered whole take many other forms. Three common types of model in particular throw light on the discontinuities, and continuities, between the philosophers' ideas and what went before. First, the cosmos was conceived as a living organism, second as an artefact, and third as a political entity.[10] In each case the idea incorporated in the model has certain mythological uses, and in each case *particular* applications of the idea appear early on in Greek philosophy. Yet *generalised* applications of the idea (the idea pressed into service to give an account of the world-order as a whole) are in two of the three cases, at least, comparatively late – to judge, that is, from the extant texts.

Take first the use of craftsman images. This idea has obvious pre-philosophical antecedents in, for example, the myth of the creation of Pandora, the first woman, by Hephaestus from earth and water in Hesiod (not to mention even earlier Egyptian and Babylonian parallels).[11] Yet in myth what is described in these terms is the creation of particular objects. Moreover, they are the work of a personal god. Hephaestus makes Pandora on Zeus' instructions to spite Prometheus, who had stolen fire and given it to mankind. In cosmology, by contrast, personalities are irrelevant, indeed excluded. We have a number of examples of craftsman images used in particular contexts in Presocratic philosophers: the notion of a force that guides or pilots or steers the world is particularly common,[12] and Empedocles especially has a wealth of craftsman images to describe the role of his cosmic force Philia (Love).[13] But the first clear extant general statement of the world as a whole as the product of a craftsmanlike agency comes in Plato: the work of the Craftsman, is depicted in vivid metaphors in the *Timaeus*, many of them being metaphors that had been used of Zeus and Hephaestus in Homer or Hesiod. But the important point in Plato is the underlying message, and this, put baldly, is that the world as a whole is the product of design.

Similar points can be made concerning the other two types of models I mentioned. The notion that the primary physical elements are alive and divine is as old as Greek philosophy itself. While we can connect the living earth and air of the early cosmologists with earlier ideas of a divine Gaia in Hesiod, for instance, the more important point is again that in cosmology earth and air, while still divine, are not personal gods. They have no will. They are left with one and only one of the properties of the living, namely the capacity for self-movement. But so far as general cosmological applications of vitalist ideas go, it is again Plato that supplies us with our first definite extant text ('this world is in truth a living creature, endowed with soul/life and reason'),[14] although vitalist notions are important much earlier in accounts of how the world developed from an undeveloped, undifferentiated state,[15] and even though the idea that the world is a living creature may already be implicit in the comparison that Anaximenes drew between the role of air in the world and that of breath in man.[16]

Thirdly, with political images, the earliest use of the model of political and social relations in cosmology may be in the very first philosophical fragment we possess, the quotation from Anaximander to the effect that certain things 'pay the penalty and recompense to one another for their injustice, according to the assessment of time.'[17] The problem of what the things in question are is disputed, but they appear to be cosmological factors of some sort, conceived as equal opposed forces. There is no question of a *conscious* selection of the political sphere to provide a model for cosmology. Rather, political, or in this case more precisely legal, terms provide the metaphors in which the idea of the stable interrelation of different things is conveyed. Again the similarity to, and contrast with, pre-philosophical beliefs are evident. In the Olympian religion, Zeus the sky-god rules over the other gods, though Poseidon, the sea, and Hades, the underworld, claim equal rights with him.[18] The notion of the gods forming a society like that of man is worked out in the fullest detail in Homer, and described so vividly that we almost forget that the gods in question are gods of the sky, the sea, fire and so on. But in philosophy the factors that are now related together are not personal deities, autocratic, touchy, unpredictable, but depersonalised factors, physical elements such as earth, air, fire and water, or opposite qualities such as the hot, the cold, the dry and the wet. And the variety of political, social and legal metaphors used in early Greek cosmology is remarkable. The notions of justice, ἰσονομία (equality), war, strife, rule, contract, are used by one Presocratic after another to convey different conceptions of how the world as we know it, made up of a variety of different things, is nevertheless an ordered whole.

The evidence suggests, I believe, that the notion of the world as a unity of interrelated parts was achieved only as a result of a good deal of speculative effort. As was to be expected, and as happens frequently in myth of course, the philosophers drew on familiar conceptions, for example of living things, of technology, of politics and society, for their ideas. But although the links with pre-philosophical beliefs are in many cases striking, the achievement of the philosophers was to arrive at *generalised* and *depersonalised* notions of the world-whole as a unity.

This takes me to my second theme, what I called the pluralism of Greek mythology. I have mentioned three common types of model: but each of these is used to express more than one cosmological doctrine. The same type of model may be, and in fact was, used to convey widely different conceptions of the world-whole. Thus the political analogies used in early Greek cosmology comprise at least three main kinds: (1) a view of the cosmos as ruled by a single supreme principle (Plato provides the chief example: in the *Philebus* (28c), for instance, Socrates is made to say that 'all wise men agree . . . that reason is king of heaven and earth'. In Plato, especially, this idea is connected with the conception of the world as an artefact: the benevolent designing agency acts now as king of a state, now as master-craftsman);[19] (2) a view of the cosmos as a balance of equal but opposed forces (for example, Anaximander, Parmenides in the *Way of Seeming*, and Empedocles);[20] and (3) a view of the cosmos as a place where war and strife are universal. Heraclitus criticised Homer for a line in which the wish was expressed that strife should be abolished from among gods and men, and he also probably implied a criticism of Anaximander when he remarked that justice *is* strife.[21] The message is that the interaction of opposites, described as war or strife, is universal: what is 'right' and 'just' and normal or usual *is* just this constant interaction of opposites. The conception of the world in political terms can, then, take many forms, depending on whether the conception is of the world as a kingdom, as an oligarchy of balanced powers, or as an anarchy, and if we were dealing with Greek political philosophy rather than with Greek cosmology, we could trace how different cosmological views of the world as a state are related in turn to different political theses or doctrines.

There is no such thing as *the* cosmological model, *the* cosmological theory, of the Greeks. One is hard put to it to describe the *predominant* notion or notions in Greek cosmology, though I shall make an attempt to identify some important themes later. Indeed, one can and must go further: one of the remarkable features of Greek cosmological thought is that for almost every idea that was put forward, the antithetical view was also proposed. For every cosmology, there is, one might say, a counter-cosmology, suggested by the Greeks themselves.

Let me illustrate.

One of the principal ideas that the model of the cosmos as an artefact was used to convey is that the world is the product of, or under the guidance of, a rational, usually benevolent, intelligence. In the fourth century BC this idea had the full weight of the authority of Plato and Aristotle, although Plato and Aristotle differ in their expression of it: for Plato the Craftsman is, or at least is described in terms that appear to imply that it is, transcendent, while Aristotle's view is of an immanent force, in that nature herself is purposeful. But the idea that the universe is the product of design had been denied, in the late fifth century, by the atomists Leucippus and Democritus, and it was so again by Epicurus and later Epicureans.[22] If teleologists often dominated the argument, anti-teleologists also had a hearing. Though little of the work of Leucippus and Democritus remains, the anti-teleological view-point is forcefully expressed in the letters of Epicurus and by Lucretius, who make clear their belief that the world is the result not of design, but of necessity, that is the mechanical interactions of atoms.

Secondly, it was, as I have remarked, commonly assumed that the physical elements are alive. Here we can trace argument and counter-argument through several generations of philosophers. Aristotle criticised his predecessors severely for believing that soul is intermingled in the whole universe, and also for the opinion that the elements themselves are alive (why does not the soul which exists in air or fire form an *animal*?).[23] Yet despite the scorn that Aristotle poured on the idea, in the fourth century and later we find the Stoics confidently reaffirming the doctrine that the cosmos as a whole is a living creature.[24]

A third example from physical theory, involving some of the thinkers that have already been mentioned, is the debate between atomism and continuum theory.[25] This dispute goes back before Aristotle, but it is after him that it becomes most vigorous. The Stoics worked out a sophisticated continuum theory of matter, picturing the world as a plenum, and conceiving movement as the transmission of a disturbance in an elastic medium, in opposition to the fourth century proponents of atomism, the Epicureans, who held that matter, space and time

all exist in the form of discrete indivisible units and who interpreted movement in terms of the transport of material particles
through a void.

A fourth and a fifth debate concern the questions of whether
the world is eternal or created, and one or many. We can find
proponents for no less than five different views: (a) the universe is one and eternal (for example, Aristotle);[26] (b) the universe is one and created (Plato, at least on one construction of
his views);[27] (c) the universe is one and alternately, and ever-
lastingly, created and destroyed (Empedocles, again according
to one interpretation);[28] (d) there are innumerable worlds that
exist (d1) in succession (a view attributed to Anaximander in
Simplicius, for example)[29] or (d2) co-existent (the atomists:
thus Democritus' pupil Metrodorus of Chios is reported to
have said that it is as unlikely for one ear of corn to be produced
in a great plain, as for one world in the boundless void).[30]

I have concentrated so far on physical and cosmological
theories, and their counter-theories, but if we broaden the
field of discussion for a moment, we find the same pattern
elsewhere. In theology, alongside the various versions of poly-
theism and what one may call, with some reservations, monotheism, both agnosticism and atheism are expressed. God does
not exist, or we cannot know whether he does or not. As the
fifth-century sophist Protagoras put it: '. . . concerning the
gods I am unable to discover whether they exist or not, or
what they are like in form; for there are many hindrances to
knowledge, the obscurity of the subject and the brevity of
human life'.[31] And although it is notoriously difficult to be
sure quite what is meant when men like Prodicus and Critias
were labelled atheists, it is obvious enough from the castigation
of those who believed that there are no gods in Plato's *Laws*
(Book X) that Plato, at least, took seriously the threat of the
denial of the existence of the gods.

Again, on the question of the sources of knowledge in
general, the dispute that begins with Heraclitus and Parmenides
continues right through the history of Greek philosophy.
Parmenides, especially, must be mentioned, for not only was

he the first philosopher explicitly to deny the validity of the senses,[32] but beginning with the statement 'it is and it cannot not be' he produced not, of course, a cosmology, but what he calls a Way of Truth, leading to the conclusion that what is is ungenerated, indestructible, unchanging, present and continuous.[33] After Parmenides' denial that change and coming-to-be take place, the status of γιγνόμενα, things that come to be, was a major problem in Greek philosophy. The central dispute in epistemology ranged round the arguments for and against reason on the one hand, and sensation on the other, and we find quite early on indications of the growth of scepticism. The weakness or limitations of human knowledge is expressed by Xenophanes: '. . . . no man knows or ever will know the clear truth about the gods and about everything I speak of . . . seeming is wrought over all things'.[34] And Democritus has this: '. . . in reality we know nothing, for truth is in the depths' (Fr 117). In another passage Democritus represents the senses saying to the mind: '. . . wretched mind, do you take your evidence from us and then throw us down? That throw is your overthrow.'[35] In the fifth century a statement of an extreme position is found in Gorgias, On What is Not or On Nature, although this is another text whose interpretation is highly disputed. It began: '. . . he says that nothing exists, but that if it exists, it is unknowable, and that if it exists and is knowable, it still cannot be indicated to others.' From the fourth century onwards we can trace a dispute, or at least differences of opinion, among the sceptics themselves, in that some denied that knowledge is possible, while others took the view that such a denial was itself a dogmatic assertion, and that as any sort of dogmatic assertion must be ruled out, the true sceptic must on this, as on every other question, withhold judgement. Some sceptics did not allow themselves to *assert* the key doctrine of scepticism, that knowledge is not possible.[36]

Many other examples could be given, but I have said enough to illustrate the theme of the pluralism of Greek cosmology. But I have so far merely indicated *what* different theorists believed, and not *why* they believed it, and I must now turn to the question of criteria and methods. In the great majority of cases, the cosmological ideas I have mentioned were worked out in direct opposition to the views of other theorists. There are two points here. First, the actual *evidence* adduced in connection with general cosmological theories down to the fourth century amounted, in most cases, to no more than a few well-known data, and the main support and justification for cosmological doctrines came from rational *argument*. Secondly, criticism of other cosmologists' views is a constant characteristic of Greek cosmological speculation. This begins, probably, very early: there is some evidence that Anaximander implies criticisms of views that can be attributed to Thales, for example. Where it seems that Thales held that the earth floats on water, Anaximander is reported to have held that it 'hangs freely' 'remaining where it is because of its equal distance from everything'.[37] But it is typical that Anaximander's view is not supported by, and was presumably not suggested by, empirical evidence of any sort. The impulse to put forward this theory came rather (one may suppose) from Anaximander's realisation of the force of the argument that Thales' view, and all views like it, runs into one obvious difficulty: if water holds the earth up, what holds the water up?[38] Throughout the Presocratic period, and despite the fact that the men concerned often inhabited widely separated parts of the Greek world, Greek philosophers were remarkably well informed about each other's ideas. The extant fragments bear witness to the impact that Parmenides' arguments had on Empedocles and Anaxagoras, as well as on his own followers Zeno and Melissus.[39] Later we can trace the disputes between the atomists and Plato, between Plato and Aristotle, between the Stoics and the Epicureans, and so on, right down to the controversies between the Christian and pagan commentators of Aristotle in the sixth century AD.

Greek cosmology is nothing if not dialectical. And this is not an accidental or contingent feature of Greek cosmology, but of the essence of the Greek contribution. What marks out philosophical cosmology from mythology is first that the former

o

proposes definite and comprehensive accounts of the world as an ordered whole, where the order does not depend on the arbitrary will of gods or divine beings, and second that the former is, in a sense that does not apply to the latter, critical. It is no part of the mythologist's purpose to produce a better, in the sense of a *truer* account than that of other myths. If he is in competition with his contemporaries and predecessors at all, it is not, generally speaking, on the matter of who produces the most rational, or best argued, or most consistent account. The cosmologists, on the other hand, *were* in competition with one another in that way. The demand is for the best explanation, the most adequate theory. They are obliged to consider the grounds for their own ideas, and the evidence and arguments in their favour, and they do their best not only to strengthen their own case but also to undermine their opponents'. The *means*, as I have said, consisted mostly of *argument*: in early Greek cosmology the evidence that was adduced, indeed the evidence that could be adduced, consisted of a few well-known facts, and in some cases the same familiar data were used on both sides of a cosmological debate. Nevertheless from the beginning there is an awareness of the need to examine and assess theories in the light of the grounds adduced for them.

As time goes on, both the range of evidence taken into account broadens – at least in certain fields – and the handling of deductive argument becomes increasingly sophisticated. One contrast between earlier and later cosmology relates to the availability of more detailed knowledge of astronomical phenomena, that is of such problem-posing facts as the irregularity of the seasons, the anomalies of the moon's movement, the stations and retrogradations of the planets. But while the range of data to hand increases, so too does the ambitiousness of the doctrines evolved to account for them. Astronomy provides one of the best, certainly the best-known, example of one of the great strengths of Greek science, the application of mathematical methods to the explanation of natural phenomena. This begins with the doctrine of concentric spheres put forward by

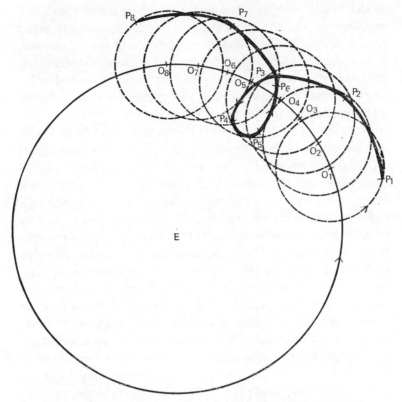

15. The epicyclic model used to explain the retrogradation of the planets

Eudoxus in the fourth century, and continues in the doctrine
that superseded it, that of epicycles and eccentrics, originally
proposed by Apollonius in the third century. The theories are
complex, too complex to attempt to describe here, although I
should remark that it is sometimes forgotten that the data these
theories were designed to explain are themselves highly complex. But the fundamental point is that the strength of both
types of model lies in the rigour of their geometry. Here too
we have the phenomenon of model and counter-model, of
modification and further modification. But from Eudoxus
onwards, while argument raged round the advantages and dis-
advantages of one model against another, that *some geometrical*

model would provide the solution to the problem of celestial motion was common ground to all Greek astronomers. After Euclid especially, the *Elements* provided a model of method, in particular a model of an axiomatic, deductive system, not only for other mathematicians and astronomers, but also for work in other physical sciences, as for example that of Archimedes in statics and hydrostatics.[40]

I have remarked how difficult it is to generalise about Greek cosmology. It is quite impossible to talk of *the* cosmological theory of the ancient Greeks. Yet if we concentrate on dominant trends, and if we stand back and compare common Greek assumptions with the assumptions that influence our own attitudes to the world we live in and man's place in it, certain broad contrasts stand out. Much of Greek cosmological thought is anthropocentric. True, the rejection of anthropomorphic gods occurs early in Greek thought. Xenophanes in the late sixth century BC remarks, in a famous fragment, that 'if oxen and horses and lions had hands and could draw with their hands and produce works of art like men, horses would draw the forms of the gods like horses, and oxen like oxen, and they would make their bodies such as each of them had themselves'.[41] Nevertheless, throughout Greek cosmological thought it remains true, as a general rule, that man's position in the universe is a privileged one. First, man is the highest member of the animal kingdom. Secondly, the earth we inhabit holds a privileged position in that it occupies the centre of the universe.

The first doctrine was never seriously questioned, and it had some strange manifestations: Aristotle, no mean biologist, asserts that man of all animals is most in accordance with nature, and that the parts of the body are, in man, in their natural position – it is natural to be upright.[42] The second theory, however, was contested. First there were those, as we have seen, who believed in innumerable worlds, although they did not necessarily deny geocentricity, especially when their innumerable worlds were successive, not co-existent. Democritus is reported to have believed that 'there are innumerable worlds of different sizes. In some there is neither sun nor moon, in others they are larger than in ours and others have more than one. These worlds are at irregular distances, more in one direction and less in another, and some are flourishing, others declining. Here they come into being, there they die, and they are destroyed by collision with one another. Some of the worlds have no animal or vegetable life nor any water.'[43] This opens the possibility of worlds being arranged in different ways, though it does not pronounce on the question of the position of the earth in our world: in fact our other evidence suggests that both Leucippus and Democritus assumed geocentricity. But then there were those who denied that the earth is at the centre of the solar system. The first to do so were some of the Pythagoreans in the late fifth century: the evidence is in parts confused, but it is reasonably certain that one Pythagorean theory (associated with the name of Philolaus) was that neither the earth nor the sun occupied the centre, which was the place of an invisible central fire. But interestingly enough, one of the reasons for displacing the earth from the centre, and it may have been the chief reason, was a religious or symbolic one: the earth is not *noble enough* to occupy the centre.[44] But while these were the first to deny geocentricity, the first astronomer to put forward an astronomical theory based on heliocentricity was Aristarchus of Samos in the third century BC. Although the evidence is indirect, it is incontestable. Archimedes, a close contemporary of Aristarchus, reports that 'Aristarchus brought out a book of certain hypotheses, in which it follows from what is assumed that the universe is many times greater than that now so called. He hypothesises that the fixed stars and the sun remain unmoved; that the earth is borne round the sun on the circumference of a circle . . . ; and that the sphere of the fixed stars, situated about the same centre as the sun, is so great that the circle in which he hypothesises that the earth revolves bears such a proportion to the distance of the fixed stars as the centre of the sphere does to its surface.'[45]

Here then was a fully fledged heliocentric theory, combining the two doctrines of the daily axial rotation of the earth and the

yearly revolution of the earth round the sun. But, as is well
known, Aristarchus' views found little favour, not merely with
the man in the street, but with other astronomers. Only one
other ancient astronomer, the Babylonian Seleucus of Seleucia,
maintained the heliocentric theory. The most important Greek
astronomers after Aristarchus, Hipparchus (second century BC)
and Ptolemy (second century AD) both denied it. The reasons
are complex, and in the case of the astronomers, at least, religion had little or nothing to do with it.[46] But although Ptolemy
agrees that axial rotation enabled the phenomena relating to the
stars to be explained, he believes that the physical and astronomical arguments against axial rotation and heliocentricity
are overwhelming.[47] First, there was the physical argument
from the observed effects of gravity on earth. Second, there
was the physical objection that the speed of the rotation of the
earth must be enormous, and yet this movement had no
observed effect on objects moving through the air. Third, the
main astronomical argument against heliocentricity was the
apparent absence of stellar parallax, the change in the relative
positions of the stars observed from the earth at different
points on the earth's orbit. And to these considerations one
may add, fourthly, that on one problem that had taxed astronomers since the mid-fourth century BC, and where quite exact
data were available, namely on the irregularity of the lengths
of the seasons, it was evident that heliocentricity by itself was
inadequate as an explanation. And finally heliocentricity was
also inadequate, and indeed irrelevant, to the solution of
the problem of the irregular movements of the moon. The
fact that in the case of the movements of the sun and the moon
eccentricity and/or epicycle motion had (it was assumed) to be
postulated must, one supposes, have weakened the attractions
of heliocentricity in the case of the movements of the planets.
Yet, so far as we can judge from Ptolemy, the main counter-
arguments to heliocentricity that weighed with him were the
two physical ones, the doctrine of natural places and the
absence of any observed effects of axial rotation. For all the
mathematical or geometrical character of Ptolemy's system

(and elsewhere he is prepared to ignore certain quite major physical objections to his theories when he believes those theories to be adequate mathematical solutions),[48] on the question of the position of the earth and whether it is at rest, he is not prepared to abandon certain fundamental physical principles, despite the increased simplicity and economy that would have resulted.

Heliocentricity was known, but for a complex of reasons, some of them, in the state of knowledge at the time apparently good ones, geocentricity was preferred. The common assumption, by laymen and astronomers alike, was that the earth is at the centre of the universe. It is interesting, next, to consider the development of an awareness of the *dimensions* of the universe. To begin with, Anaximander appears to have believed that the sun is further away than the stars. The sun is at twenty-seven earth-diameters, the moon at eighteen and the stars at nine from the earth, the earth itself being pictured as a flat-topped cylinder three times as wide as it is deep.[49] Here the multiples of three give the game away: it is *symmetry* that counts, and the dimensions in question are not the result of observation or measurement. The second stage is represented by Aristotle. He records the first precise estimate of the dimensions of the earth (the figure he gives for the circumference of the earth is almost twice the actual one) and he also remarks that 'the bulk of the earth is as it were nothing compared with the surrounding universe'.[50] A *general* idea that the earth is tiny compared with the heavens thereafter becomes a commonplace, and from the fourth century BC a good deal of attention was paid by astronomers to the topic of the sizes and distances of the sun and moon. The nature of their work on this problem is, however, rather different from what we might expect. One astronomer who tackled the subject is Aristarchus. His treatise *On the Sizes and Distances of the Sun and Moon* is the one complete work of his that is extant. It is a fascinating document that raises many thorny problems of interpretation, not least among them why Aristarchus should have assumed a figure of two degrees for the angular diameter of the moon, when the fact that the angular diameter is approximately one-half a degree was well known to the Greeks before Aristarchus, and when indeed we are told by some of our sources that Aristarchus himself adopted that value. Why, then, should he have chosen a figure so wide of the mark in his treatise *On the Sizes and Distances*? Various explanations have been put forward, but part of the answer, at least, lies, I believe, in the point that in that treatise, despite its title, Aristarchus is interested not so much in arriving at concrete results – figures for the sizes and distances in question – as in the solution of the geometrical problems that the topic posed. He was not interested, that is, in setting out his estimates, in stades, for the values in question: in fact there are no concrete results set out in *that* form in the work at all, since the conclusions of the work are expressed as proportions, stating the *relations* between different diameters and distances.[51] His chief concern appears to be in the mathematical exercise, the solution of the geometrical problem, and for this purpose the accuracy of the figure he assumed for the diameter of the moon was unimportant.

It would, however, be wrong to leave the impression that Greek astronomy is merely theoretical, or merely interested in the solution of mathematical problems, although that is undoubtedly its forte. Important work was also done in observational astronomy, and some of this is relevant to the world-picture of the scientist, if not to that of the man in the street. Of course, Greek observational astronomy owes an enormous debt to the Babylonians. Ptolemy, for example, draws on Babylonian eclipse records going back to the eighth century BC, and Hipparchus had probably done the same before him. But it is important not to exaggerate the contrast between Babylonian and Greek astronomy at this point, by representing the Babylonians as observers while the Greeks were theorists. At least one must recognise the solid observational achievements of the Greeks themselves. One of the more striking discoveries that resulted from detailed observation, and that was relevant to cosmology, was Hipparchus' discovery of the phenomenon

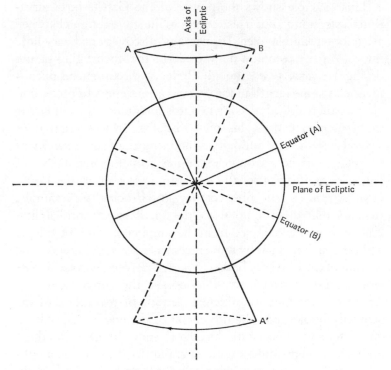

16. The precession of the equinoxes.

known as the precession of the equinoxes. The positions of the
equinoctial points, defined as the intersection of the ecliptic
and the celestial equator, do not remain constant in relation to
the fixed stars, but are displaced from east to west very slightly
each year. Indeed, Hipparchus not only detected this phenomenon by comparing his observations with those of Greek
astronomers working some 160 years earlier, but he also gave
an astonishingly accurate estimate of the *rate* of precession. A
passage in Ptolemy[52] implies that he set a figure of 1° in a
hundred years as the lower limit, but the actual figure he
adopted may have been 2° in 160 years, or 45 seconds of angle a
year, within six seconds of that determined by modern
astronomers.

This example shows the power of the Greeks in observational astronomy, but it also serves to illustrate the gap between astronomers and laymen. Hipparchus' discovery had very little impact on the common notions about the unchanging nature of the heavens. Even though Ptolemy discusses the phenomenon at some length in the *Syntaxis*, references to precession are rare thereafter. Indeed, when it is mentioned, it is often for the writer to express his own disbelief. It was referred to several times in late antiquity as an instance of the *absurdity* of the views of the astronomers, along with the hypotheses of epicycles and eccentrics that constitute the core of the Ptolemaic system. In the fifth century AD, Proclus, for example, remarks that these hypotheses 'do not have any probability, but some are far removed from the simplicity of divine things, and others, fabricated by more recent astronomers, suppose the motion of the heavenly bodies to be as if driven by a machine'.[53] And in the sixth century, Philoponus, the commentator on Aristotle, mentions the different periods of revolution of the heavenly bodies, and the precession of the equinoxes, only to go on to say: 'who would be able to state the cause of these things? . . . This only we can say, that God has made everything well and as is needed, neither more nor less.'[54] Each in his own day, Proclus and Philoponus were among the foremost theorists who were interested in the investigation into nature. If one turns to other writers, particularly among the Christian fathers, it is sometimes not just the precession of the equinoxes that is flatly disbelieved, but also the sphericity of the earth.[55] The discovery of precession shows, then, the problem of the lack of penetration of scientific ideas: it must be remembered that the numbers of those who progressed beyond an elementary knowledge of mathematics, astronomy and biology were always – even in the heyday of Greek science in the third century BC – very small. And this example also illustrates the problem of the loss of knowledge during the decline of Greek science, although that is beyond my brief in this paper.

The history of early Greek cosmology is one of argument and counter-argument with a paucity of references to empirical data, and those mostly familiar ones. Yet in the natural sciences, if not in cosmology as such, the use of, on the one hand, deductive methods, and, on the other, of observation, measurement, even, on occasion, controlled experiment, was, in the hands of certain scientists at least, highly developed. The Greek speculative imagination was extraordinarily fertile in ideas, some of them strongly suggestive of a point of view opposed to the prevailing anthropocentric emphasis, such ideas as heliocentricity, innumerable worlds, and the precession of the equinoxes. Nevertheless, despite the fertility in ideas, and despite the development of criteria and methods, the dominant cosmological view remained anthropocentric. The victory of geocentricity over heliocentricity was both a symptom and a cause of this. The observations of the biologists meanwhile (which included the use of human dissection and vivisection in Alexandria in the third century BC, and the extended use of experimental animal dissections and vivisections by Galen in the second century AD)[56] were often used to confirm the proposition of the privileged place of man among the animals. Notions of the development of natural species, or at least of changes in existing species, the survival of the fit, and the extinction of the unfit, the development of man from a primitive to a more civilised state, can be traced back into the Presocratic period;[57] but the prevailing view was that of the fixity of natural species. Moreover, whether one believed in fixity or change, it was common ground to both sides of the argument that man is supreme among the animals.

The anthropocentrism of Greek cosmology and science is in certain respects at least, a weakness. Harsher critics would say a symptom of a failure of nerve. Yet from another point of view this characteristic of the dominant (but, I repeat, not the only) strain in Greek cosmological thinking has, perhaps, a moral for us today. We live in a world that is far stranger than anything the Greeks dreamed of. If the names of astronomy and physics have not changed, their contents have, almost totally. Yet while, at the frontiers of knowledge, the pure scientist is engaged in an activity which, while it is quite different in content, yet remains similar in spirit to that of the ancients, the role of science in society, and the expectations that society entertains of science, are nowadays very different from the ancient situation. The notions of applying knowledge to practical ends, and of controlling nature, can be found in ancient writers, despite what has been written by some modern critics. But even among those ancient writers who are aware of the possibility of applying science to practical ends, the chief motive for the investigation of nature generally remains the non-practical one. The ancients explored nature not to dominate her, certainly not to exploit her, but to become wise. The aim was understanding: and the view was often expressed that without understanding and knowledge, peace of mind and happiness are unattainable. Science had no need to be uneasily defensive in relation to a society, or to governments, that demanded concrete material benefits in return for a massive expenditure. There was simply no massive expenditure, and the scientist felt no more obliged to defend or justify his activities than did the poet or the philosopher. Science was, indeed, a part of philosophy, and it was treated by some ancient writers as an aid to improving the character. Ptolemy has this to say about astronomy: '. . . of all studies this one especially would prepare men to be perceptive of nobility both of action and of character: when the sameness, good order, proportion and freedom from arrogance of divine things are being contemplated, this study makes those who follow it lovers of this divine beauty, and instils, and as it were makes natural, the same condition in their soul.'[58] And again, in an epigram attributed to him: 'I know that I am mortal, a creature of a day: but when I search with my mind into the multitudinous revolving spirals of the stars, my feet no longer touch the earth, but beside Zeus himself, I take my fill of ambrosia the food of the gods.'[59] We may smile at the idea of Ptolemy supping with the gods: but we cannot, I think, fail to be struck by the ideal of science as an inquiry to be pursued as an end in itself, as part of the good life, and indeed of moral education. Greek cosmology and science are anthropocentric. But the other side of that coin is that they are also confidently humanist, confident that the primary justification of the inquiry is not practical applicability, but knowledge. Science should be useful, but the criterion of usefulness is not material welfare, but understanding.

Notes

1. This is printed here as delivered in lecture form except for some minor alterations of phraseology and slight additions.
2. I have discussed the problems from one particular point of view in *Polarity and Analogy* (Cambridge, 1966).
3. See *Iliad* 14, 201 and 246.
4. *Theogony* 108 ff.
5. *Theogony* 116 ff.
6. See, for example, Aristotle, *Metaphysics* 1091 b 4 ff., Aristophanes *Birds* 693 ff., Damascius, *de principiis* 123 ff. The evidence is evaluated in G. S. Kirk and J. E. Raven, *The Presocratic Philosophers* (Cambridge, 1957), Ch 1.
7. *Theogony* 129, 381 f.
8. See, for example, Euripides Fr. 484 Nauck.
9. There is a full analysis of the evidence for Milesian cosmology in W. K. C. Guthrie, *A History of Greek Philosophy*, Vol. 1 (Cambridge, 1962), Ch. 3. For a brief statement of one interpretation, see Ch. 2 of my *Early Greek Science, Thales to Aristotle* (London, 1970). Another sharply differing view has recently been put forward by M. C. Stokes in *One and Many in Presocratic Philosophy* (Center for Hellenic Studies, Washington D.C., 1971).
10. These three types of models are discussed in some detail in Ch. 4 of *Polarity and Analogy*. Cf. especially G. Vlastos, 'Isonomia', *American Journal of Philology*, 74 (1953), 337–66, and F. Solmsen, 'Nature as Craftsman in Greek Thought', *Journal of the History of Ideas*, 24 (1963), 473–96.
11. *Theogony* 570 ff., *Works* 59 ff.
12. See Aristotle, *Physics* 203 b 10 ff., Heraclitus Frr 41 and 64, Parmenides Fr 12, Diogenes of Apollonia Fr 5.
13. For example, Frr 73, 87 and 96.
14. *Timaeus* 30b.
15. This idea was probably already present in Anaximander: see ps.-Plutarch, *Strom.* 2. Some of the Pythagoreans are reported as believing

that the original 'one' from which the complex cosmos developed was composed of seed, see, for example, Aristotle, *Metaphysics* 1091 a 15 ff.

16. Aetius I, 3, 4 ('fragment' 2).

17. Fr 1, from Simplicius, *in Ph.*, 24, 19 ff.

18. *Iliad* 15, 185 ff.

19. Cf. also Heraclitus Fr 53, Anaxagoras Fr 12 and Diogenes of Apollonia Fr 5.

20. See Anaximander Fr 1, Parmenides Fr 9 4 and Empedocles Fr 17 20 and 27 ff.

21. Aristotle, *Eudemian Ethics*, 1235, a, 25 ff., and Heraclitus Fr 80 ('war is common and justice is strife and everything happens through strife and necessity').

22. See, for example, Aristotle, *Physics* 196 a 24 ff., *GA* 789 b 2 ff., Epicurus, *Letter to Herodotus*, 76 f., *Letter to Pythocles*, 88 ff.

23. *De Anima* 411 a 9 ff.

24. For example, Sextus Empiricus, *Against the Physicists*, I 104 (on Zeno), Diogenes Laertius VII 142 ff. (on Chrysippus).

25. The evidence is collected and discussed in S. Sambursky, *The Physics of the Stoics* (London, 1959), and 'Atomism versus continuum theory in ancient Greece', *Scientia*, 96 (1961), 376–81.

26. Aristotle presents arguments for the unity and eternity of the world in, for example, *De Caelo* I Chs 8–12.

27. As is well known, the question of whether Plato intended his account of the coming-to-be of the cosmos in the *Timaeus* to be taken literally, or whether he gave his account that form merely for the sake of convenience of exposition, is one that was disputed already by Plato's own immediate successors in the Academy. See Aristotle, *De Caelo* 279 b 17 ff., 32 ff.

28. See, for example, Fr 17. The details of Empedocles' cosmology have been subject to widely differing interpretations. See especially U. Hölscher, 'Weltzeiten und Lebenszyklus', *Hermes*, 93 (1965), 7–33; F. Solmsen, 'Love and Strife in Empedocles' cosmology', *Phronesis*, 10 (1965), 109–48; J. Bollack, *Empédocle*, 3 vols in 4 (Paris, 1965–9); and D. O'Brien, *Empedocles' Cosmic Cycle* (Cambridge, 1969).

29. See especially the Additional Note, pp. 106–15 in Vol. 1 of Guthrie, *A History of Greek Philosophy* (Cambridge, 1962).

30. Aetius I 5 4. Cf. Diogenes Laertius IX 31, Epicurus, *Letter to Pythocles* 89.

31. Fr 4 quoted, for example, by Diogenes Laertius IX 51.

32. Fr 7.

33. Parmenides goes on to present a cosmology in the *Way of Seeming*, but he makes it clear that he does not consider this account to be true. At Fr 8 52 ff., introducing it, he says, 'listen to the deceitful ordering of

my words', even though he includes the account 'so that no judgement of mortals shall outstrip you' (V 61).

34. Fr 34.

35. Fr 125.

36. See, for example, Sextus Empiricus, *Outlines of Pyrrhonism* I, for example, 236–41. For a recent account of the history of scepticism, see C. L. Stough, *Greek Skepticism, A study in epistemology* (Berkeley and Los Angeles, 1969).

37. Aristotle, *De Caelo* 294 a 28 ff., 295 b 10 ff., Hippolytus, *Ref.*, I 6 3.

38. Cf. the argument for postulating the Boundless as ἀρχή reported by Aristotle at *Physics* 204 b 24 ff.

39. See Empedocles Frr 12–14, Anaxagoras Fr 17.

40. There is a brief account of the theories mentioned in this paragraph in Ch. 7 of *Early Greek Science* and Chs 4–5 of *Greek Science after Aristotle* (London, 1973).

41. Fr 15.

42. For example, *IA* 706 a 19 ff., *PA* 656 a 10 ff.

43. Hippolytus, *Ref.*, I 13 2, translation from Guthrie, *A History of Greek Philosophy*, Vol. 2 (Cambridge, 1965), p. 405.

44. See Aristotle, *De Caelo* 293 a 21 ff., 30 ff., b 23 ff.

45. Archimedes, *The Sandreckoner*, introduction (II, 244, 9 ff., Heiberg).

46. We hear, however, from Plutarch (*On the face of the moon*, Ch. 6, 923a) that the Stoic philosopher Cleanthes 'thought that the Greeks ought to indict Aristarchus of Samos on a charge of impiety for putting in motion the Hearth of the Universe [that is, the earth]'.

47. See *Syntaxis*, Book I, Chs 5 and 7. He remarks, however, concerning the hypothesis of the axial rotation of the earth, that 'so far as the phenomena relating to the stars are concerned, perhaps nothing might prevent things from being in accordance with the simpler [form of this] theory' (Book I, Ch. 7, I, 24, 14 ff., Heiberg).

48. The classic example of this is in his theory of the moon, where the values he assigned to the radii of the moon's epicycle and deferent circle had the consequence that the angular diameter of the moon should appear to vary by nearly a factor of 2. See *Greek Science after Aristotle*, 127 f.

49. The evidence is, however, fragmentary, and its interpretation highly disputed. See, for example, C. H. Kahn, *Anaximander and the origins of Greek cosmology* (New York, 1960), 58–63; Guthrie, *A History of Greek Philosophy*, Vol. 1, 93 ff., and D. R. Dicks, *Early Greek Astronomy to Aristotle* (London, 1970), 45 f.

50. *De Caelo* 298 a 6–20. *Meteorologica* 340 a 6 ff., cf. 352 a 27 f.,

51. For example, 'the distance of the sun from the earth is greater than 18 times, but less than 20 times, the distance of the moon from the earth' (Proposition 7). T. L. Heath, following Hultsch and others, gives an account of the later history of the problem in *Aristarchus of Samos*

(Oxford, 1913), Part 2, Ch. 4. It should be noted that despite the title of this chapter in Heath ('later improvements on Aristarchus' calculations') it was not always the case that later astronomers improved on their predecessors' results.

52. *Syntaxis* Book VII Ch. 2, II, 12, 21 ff., Heiberg.

53. *In Ti.* III, 56, 28 ff., Diehl.

54. *De opificio mundi* III 4, 117, 15 ff., Reichardt.

55. An extreme example is provided by the fantastic cosmology of the *Christian Topography* of Cosmas Indicopleustes of Alexandria in the sixth century, but Lactantius, too, in the fourth, had rejected the sphericity of the earth (*Divinae Institutiones* III Ch. 24).

56. The evidence (e.g. from Celsus, *De Medicina*, introduction 23 ff.) concerning the use of human dissection and vivisection by Herophilus and Erasistratus in the third century BC has, however, been disputed. See, for example, L. Edelstein, 'The History of Anatomy in Antiquity' in *Ancient Medicine* (Baltimore 1967), 247–301, and F. Kudlien, Pauly-Wissowa, Suppl. Bd. XI (1968), col. 38–48. Among the most remarkable of Galen's vivisections of animals are those in which he investigated the effects of severing one side, or the whole, of the spinal cord at different points on the spinal column, described in his *On Anatomical Procedures* Book IX, Chs 13 ff.

57. The most important evidence relates to Anaximander (Hippolytus, *Ref.*, I, 6, 6, Plutarch, *Quaest. Conv.* VIII, 8, 4, 730 ef, ps.-Plutarch, *Strom.*, 2, and Aetius V, 19, 4) and Empedocles (Frr 57 and 61, Aristotle, *Physics* 198 b 29 ff., Simplicius, *in Ph.* 371, 30 ff.). Cf. also Diodorus I, 7, 1 ff., 8, 1 ff.

58. *Syntaxis*, Book I, Ch. 1, I, 7, 17 ff., Heiberg.

59. *Anthol.* IX, 577.

9
The European Heritage

PHILIP GRIERSON

Professor of Numismatics, University of Cambridge

In the 1940s and early 1950s the late Professor C. S. Lewis, while still at Oxford, gave over a number of years the course of lectures which was published in 1964 as *The Discarded Image*. The lectures were addressed to the English faculty, and their purpose was to provide students of medieval and renaissance literature with an introduction to the cosmological assumptions of the writers with whom they were concerned: the structure of the universe, the relationship between heaven and earth, the place in the latter of man and of the animal and vegetable creation, the nature of man himself. The title, a singularly happy one, admitted that this world picture, despite its very considerable intellectual content and the greatness of the literature and art it had inspired, no longer fitted the facts and had consequently been abandoned by most people in the western world. His own attitude to the discarded image was, as he himself allowed, one of nostalgia. The structure was a beautiful one, even if one could no longer feel quite at home in it.

The theme of this chapter is much the same as that of Professor Lewis' book, but the treatment is necessarily different. Partly, it is a question of scale: a chapter is not a book. Partly, it is one of temperament: one's wish to understand the world of the past can be unaccompanied by the slightest desire to inhabit it. Partly, it arises out of a difference in emphasis: he was concerned with cosmology as a background to literature,

P

and his authors naturally made more use of some aspects than of others, though these might be of greater historical or scientific importance. Partly, it is that while he went to great pains to discuss the late classical sources and the contribution of Graeco-Roman antiquity to the world picture of medieval times, he took the biblical background, Jewish and early Christian, more or less for granted. This, to some degree, falsifies the proportions; one can in any case no longer take for granted the intimate acquaintance with the Bible on which he felt able to count only twenty years ago. Professor Lewis was also, as he frankly admitted, not greatly interested in the ways in which the picture changed and developed in the course of the Middle Ages or in the reasons for which it gradually became unacceptable to educated men.

At this point, it will be as well to emphasise the overwhelming role of religion in the world picture of medieval man. Half a century ago, a distinguished American publicist wrote of revealed religion in terms as trenchant as they were disrespectful:

'Its dogma, as Mr Santayana once said, is insensibly understood to be nothing but myth, its miracles nothing but legend, its sacraments mere symbols, its bible pure literature, its liturgy just poetry, its hierarchy an administrative convenience, its ethics an historical accident, and its whole function simply to lend a warm mystical aureole to human culture and ignorance. The modern man does not take his religion as a real account of the constitution, the government, the history, and the actual destiny of the universe. With rare exceptions his ancestors did.[2]

Whatever one may think of other aspects of this judgement, in the field of cosmology it is wholly true. Whether medieval cosmology is better regarded as theocentric, as theologians and most medievalists would hold, or anthropocentric, as in many respects it certainly was, is largely a matter of words; it was in either case entirely different from our own. Many concepts which in antiquity had become largely secular in character,

and which are now looked on as completely so, had in the
Middle Ages been reintegrated into a religious framework and
have to be studied in such a setting.

Exactly what such a term as 'the European heritage', can be
expected to cover in the field of cosmology is hard to say, but
one may take it to be basically the ideas of Latin Christendom
from the end of the Roman Empire in the West to some time
between the sixteenth and the twentieth centuries. The terminal
date must depend upon whether one regards descriptive and
analytical astronomy as the essential element in cosmology (one
can then, according to taste, end with Copernicus or Kepler or
Galileo or Newton) or whether one extends the concept to
include the origins of the universe and the role of man in it,
problems upon which medieval solutions have been very widely
abandoned, without any generally accepted alternatives having
yet emerged to take their place. Astronomers, one is given to
understand, are still divided between 'steady-state' and 'big-
bang' hypotheses, and a distinguished Princeton cosmologist
was recently quoted in *The New Scientist* (29 July 1971, p. 242)
as having described his field of study as the chemistry of
geometry in superspace, where neither time nor space exist,
but which is the true reality. An oracular pronouncement of
this nature is no encouragement to the historian to meddle in
very recent times, and it will be as well to limit ourselves to
the history and fate of what can fairly be regarded as the inheritance of antiquity.

This inheritance was of a double character, Greek and Jewish,
but each was in certain respects modified by developments in
early Christian theology. There are to our eyes inherent contradictions between the two traditions, but these contradictions
were minimised by the fact that Greeks and Jews were mainly
preoccupied with different things. The Greeks were profoundly
interested in the structure of the universe, but comparatively
little in its origins; the Jews, on the other hand, had some traditional views on its origins – it would be an exaggeration to say
that they were profoundly interested in it – while they cared

very little about its structure. It was thus possible to arrive at a workable partnership between the two.

Such complications as there were, arose out of three things. First, the Jewish tradition was that of a creation in time, the universe beginning at some moment in the past and then proceeding, under divine guidance, to some future end. Its concept of time was essentially linear, while the Greeks, preoccupied with mathematical and particularly with geometrical figures, and convinced of the superiority of circles (whose circumferences have neither beginnings nor ends) over straight lines (which in a universe that did not recognise infinity necessarily had both), were more given to conceiving of the universe as both eternal and cyclical, with an endlessly repeating pattern of renewals and breakdowns.[3] Second, the Jewish account was authoritative, set out in Scriptures that were divinely inspired, and however much Jewish thinkers might explain and embroider, there remained certain limits beyond which they could not pass. Greek thought, on the other hand, was unauthoritative; there was not one cosmological view, but several. Explanations put forward by even the most venerated of philosophers, could only be regarded as acceptable because they accorded with reason, and since these philosophers, however convinced themselves of the wholly logical character of their reasoning, disagreed on some of their premises and many of their conclusions, the ordinary man was left in a state of uncertainty, free to pick and choose between the solutions offered him or combine them in any way he pleased. Third, there were the complications introduced into the Jewish construction by Christian theology. The book of Genesis, with the authority of divine revelation and tradition behind it, provided an account of the Creation and the Fall of Man, but no scheme of redemption. Pauline Christianity, in adding the latter, seemed to make the efficacy of Christ's death dependent upon the historicity of at least the Fall.

To conceive of the Middle Ages as the simple heir to Greek and Jewish thought, to the astronomical insights of the late classical and Hellenistic periods on the one hand and to Biblical traditions on the other, would be greatly to oversimplify the picture. It is true, that so far as a knowledge of Jewish traditions outside the Bible text is concerned, it is not far wrong. Josephus was indeed accessible, but he is more important for history than for cosmology; Philo was not available in Latin and could only affect Western thought through his influence on some of the Greek Fathers; and it was not till the later Middle Ages and Renaissance that Latin thinkers were exposed to the erudition of the Talmud and the esoteric extravagances of the Kabbalah. But the great body of Greek learning was transmitted in a manner so complex and imperfect as to deprive it of much of its range and value. Three obstacles affected both Greek East and Latin West; a fourth only the Latin world at the beginning of the Middle Ages but the Greek world later on; and a fifth the Latin world only. The first three obstacles were the transition from papyrus to vellum, the prejudices of Neo-Platonism and Christianity, and the widespread taste for epitomes and encyclopedias; the fourth was loss by war or neglect; and the fifth was the problem of language, the making available in Latin of works written originally in Greek.

The first and second largely go together. The literary and scientific works of antiquity were customarily written on rolls of relatively fragile papyrus, those of the Middle Ages, in both East and West, in codices of much more durable vellum. The transition from one to the other was effected during the first centuries of our era, mainly between the fourth and the sixth, and, leaving aside what has become available in modern times from the rubbish dumps of the Fayoum and similar localities, virtually nothing that in this period was not copied from papyrus to vellum has survived at all. What was copied, however, was only a small proportion of what once existed, and would be in the main what men of that time found worthy of approval on scientific, aesthetic, or religious grounds. It is a little as if, some hundred years from now, all our books had disappeared and only microfilms were left, but with the important difference that microfilming is cheap and vellum books were dear, so that with vellum a much smaller proportion of what was worth preserving was transferred from one medium to the other. Religious and philosophical predilections also played a considerable role in determining what was copied from papyrus, so that strong differentials operated in favour of some types of work rather than others.

Loss by war and neglect chiefly affected the West, which in the course of the fifth century was conquered and occupied by various Germanic peoples who set up kingdoms of their own. It has in recent years, been fashionable amongst historians to pretend that the fall of Rome never happened. The culture of the early Middle Ages is interpreted as a natural development of trends already apparent under the later Empire; late Roman rural society is supposed to have merged insensibly into that of the various Germanic states; the great magnates of Frankish and Lombard times are thought of as playing a role not greatly different from that of the provincial aristocracy of fifth-century Gaul and Italy. Such an explaining away of the Dark Ages seems to me totally false. There was, in the intellectual sphere, a staggering decline in resources. The towns of the later Roman Empire were no doubt usually smaller and less flourishing than they had been in earlier times, but they were still well provided with libraries, while wealthy senators like Symmachus and Sidonius Apollinaris must have had private collections of many hundreds of volumes. In the confusion and chaos of the invasions, these city libraries and all the great private libraries perished; by the time we reach the Carolingian period not a trace of them remains. The episcopal and monastic libraries that replaced them were on an altogether different scale, and the achievements of scholars like Isidore and Bede are all the more admirable when we take account of the inadequacy of the material with which they had to work.

There was, in any case, the problem of language. The scientific literature of the Roman Empire had been almost entirely in Greek. During the first centuries of our era this had not greatly mattered, at least from a short-term point of view. Any person of education was likely to have a sufficient knowledge of the language, and for those who had not there was a strong encyclopedist tradition in Latin which made much of Greek science available in the West, though often in a simplified and vulgarised form. One hesitates to speak disrespectfully of Pliny, to whose immense *Historia Naturalis* every student of ancient science owes so much, but he had the defects of writers of his class. His appetite for facts was insatiable, though they had usually to be facts – or fictions – he could read in the pages of others and did not have to observe for himself, and he admitted his lack of interest in intellectual constructs, and consequently in much of the cosmological thinking of the Greeks.[4] On a long-term view, the reliance on Greek was in any case a misfortune; it militated against the emergence in the West of any centre of intellectual activity comparable to Alexandria or Athens, and its drawbacks became more and more apparent as East and West began to draw apart, from the late fourth century onwards, and the knowledge of Greek in the West began to die out. It is broadly true to say that from the sixth to the fifteenth century Greek learning had to be translated into Latin before it could be used in the West.

The main periods of translation were three, one between the fourth and the sixth centuries, the second in the twelfth and thirteenth, and the third in the fifteenth. The first was mainly concerned to make available to the Latin world the Bible and the writings of the Greek Fathers, but it incidentally saw the appearance of Chalcidius' translation of, and commentary on, most of the *Timaeus* (capp. 1–53),[5] a work of great cosmological importance and the only dialogue of Plato known in the West prior to the twelfth century. The third is mainly remembered because it first introduced the masterpieces of classical Greek literature to western Europe, though it also saw the translation of much Neo-Platonic philosophy and of the Hermetic Corpus, both of which profoundly influenced the world-picture of the Renaissance. The second period is the one that, scientifically, was really crucial. It involved at different times direct translations from the Greek, indirect translations from Greek by way of Arabic, often with extensive Arabic commentaries added, and translations at third-hand from Greek via Hebrew and

Arabic, likewise with commentaries acquired en route. The main centres were Spain (Toledo), the Norman kingdom of Sicily (Palermo), and after 1204 the Frankish states set up on the ruins of the Byzantine Empire. The works involved were mainly those of a logical, philosophical, and mathematical character, biology and medicine taking second place and literature being entirely neglected. It was through the work of the great translators, from Adelard of Bath in the early twelfth century to William of Moerbeke, Archbishop of Corinth, in the mid-thirteenth, that the vast bulk of Greek scientific writing, or rather of as much of it as either still survived at Byzantium or had been translated into Arabic, was made available to the West.

Apart from the *Timaeus*, therefore, the Latin world had virtually no direct access to Greek cosmological thought before the twelfth century. It knew it indirectly, however, at least from the Carolingian Renaissance onwards,[6] through the Latin encyclopedists, from Pliny in the first century AD to Isidore of Seville[7] in the seventh, and through two remarkable works of the early fifth century, the *Commentary on the Dream of Scipio* of Macrobius[8] and the oddly styled *Marriage of Mercury and Philology* of Martianus Capella.[9] The authors of both were pagans, but their Neo-Platonism had many elements in common with Christian philosophy and they in fact dominated Latin cosmological thought throughout the Dark Ages. Even when, in the twelfth century, their science began to look distinctly out of date, the new corpus of astronomical information which was then becoming available remained a matter for specialists and to the ordinary scholar did not seem wholly incompatible with the framework they had provided. They owed much of their success to the fact that both were works of literature, and they share one major weakness with the epitomisers and encyclopedists of the classical tradition. These writers all make a great parade of learning, citing a host of authorities for their opinions, and it can frequently be shown that not one of these was really consulted. The modern scholar, perhaps made wise by perceiving similar tendencies in himself or his colleagues, can usually identify the previous writer, conspicuously absent from the list of authorities, who supplied most of their information.

Of the two works in question, Macrobius' *Commentary* is the more agreeable reading. The *Somnium Scipionis*, which ended the last book of Cicero's lost *De Republica*, was a literary device of the same kind as the myth of Er in Plato's *Republic*. It depicted the younger Scipio, on a visit to Massinissa, caught up in a dream to the uppermost part of the heavens, where his future career was revealed to him and from where he could observe the structure of the universe and the wheeling spheres of the planets. Cicero's few but marvellously poetic pages allow Macrobius in his commentary to display his own cosmographical knowledge, as extensive as it was superficial, and to set forth his views on the nature of the soul and of justice. Martianus Capella's *Marriage of Mercury and Philology* has much less literary merit. It is an immense allegorical work, in highly affected and difficult Latin, which describes how Mercury, God of Eloquence, looking for a bride, has the *doctissima virgo Philologia* brought to his attention. When she ascends to heaven she brings with her as handmaids the seven liberal arts, Grammar, Dialectic, Rhetoric, Geometry (including Geography), Arithmetic, Astronomy and Music. Each handmaid describes the range of her knowledge at enormous length and with little sense of proportion. The range of authorities cited is highly impressive – Dicearchus, Archimedes, Anaxagoras, Pytheas, Eratosthenes, Ptolemy and Artemiodorus for geography, Eratosthenes, Ptolemy, Hipparchus, Pythagoras and Archimedes for astronomy – but Martianus knows little of their precise contributions to the subject and one suspects that he had never read a line of any of them. The geographical material comes directly or indirectly from Solinus and Pliny, whose names are omitted – Solinus in turn, it may be added, had plagiarised Pliny and Pomponius Mela, but mentions neither – and the astronomical information probably from Varro, Pliny's great predecessor whose works are unfortunately mostly lost. But the defects of Macrobius and Martianus

Capella were, given the circumstances of the time, far outweighed by their merits, and it is easy to understand how they
came to provide the world picture of the Middle Ages.

The elements of this picture are quite different from those of
today. One may examine it from several points of view: its
structure, its functioning, and its history and purpose. The
structure is essentially Greek, and one cannot do better than
describe it in terms of Scipio's dream, which emphasises the
immensity and majesty of the universe as a whole and the
insignificance in it of the earth. The elder Scipio points out to
the narrator, his younger namesake, how by a life of public
service the statesman[10]

'obtains his passport into the sky, to a union with those who
have finished their lives on earth and who, upon being released
from their bodies, inhabit that place at which you are now
looking (it was a circle of surpassing brilliance gleaming out
amidst the blazing stars), which takes its name, the Milky
Way, from the Greek word.

As I looked out from this spot, everything appeared splendid and wonderful. Some stars were visible which we never
see from this region, and all were of a magnitude far greater
than we had imagined. From here the earth appeared so small
that I was ashamed of our empire which is, so to speak, but a
point on its surface.

'As I gazed rather intently at the earth my grandfather said:
"How long will your thoughts continue to dwell upon the
earth? Do you not behold the regions to which you have
come? The whole universe is comprised of nine circles, or
rather spheres. The outermost of these is the celestial sphere,
embracing all the rest, itself the supreme god, confining and
containing all the other spheres. In it are fixed the eternally
revolving movements of the stars. Beneath it are the seven
underlying spheres, which revolve in an opposite direction to
that of the celestial sphere. One of these spheres belongs to
that planet which on earth is called Saturn. Below it is that

brilliant orb, propitious and helpful to the human race, called
Jupiter. Next comes the ruddy one, which you call Mars,
dreaded on earth. Next, and occupying almost the middle
region, comes the sun, leader, chief, and regulator of the other
lights, mind and moderator of the universe, of such magnitude
that it fills all with its radiance. The sun's companions, so to
speak, each in its own sphere, follow – the one Venus, the
other Mercury – and in the lowest sphere the moon, kindled
by the rays of the sun, revolves. Below the moon all is mortal
and transitory, with the exception of the souls bestowed upon
the human race by the benevolence of the gods. Above the
moon all things are eternal. Now in the centre, the ninth of
the spheres, is the earth, never moving and at the bottom".'

There follows an account of the music of the spheres, but
Scipio keeps turning his gaze back to earth, which provides
Cicero with an occasion for warning against pride and which
Macrobius was to take as his text for an excursus on the figure
of the earth.

'My grandfather then continued: "Again I see you gazing at
the region and abode of mortals. If it seems as small to you
as it really is, why not fix your attention upon the heavens
and contemn what is mortal? Can you expect any fame from
these men, or glory that is worth seeking? You see, Scipio,
that the inhabited portions on earth are widely separated and
narrow, and that vast wastes lie between these inhabited
spots, as we might call them; the earth's inhabitants are so cut
off that there can be no communication among different
groups; moreover, some nations stand obliquely, some trans-
versely to you, and some even stand directly opposite you;
from these, of course, you can expect no fame. You can also
discern certain belts that appear to encircle the earth; you
observe that the two which are farthest apart and lie under
the poles of the heavens are stiff with cold, whereas the belt
in the middle, the greatest one, is scorched with the heat of
the sun. The two remaining belts are habitable; one, the
southern, is inhabited by men who plant their feet in the

opposite direction to yours and have nothing to do with your people; the other, the northern, is inhabited by the Romans. But look closely, see how small is the portion allotted to you! The whole of the portion that you inhabit is narrow at the top and broad at the sides and is in truth a small island encircled by that sea which you call the Atlantic, the Great Sea, or Ocean. But you can see how small it is despite its name! Has your name or that of any Roman been able to pass beyond the Caucasus, which you see over here, or to cross the Ganges over yonder? And these are civilised lands in the known quarter of the globe. But who will ever hear of your name in the remaining portions of the globe? With these excluded, you surely see what narrow confines bound your ambitions. And how long will those who praise us now continue to do so?"'

This in turn, is followed by a passage on time – the length of the 'great year' over which all the constellations and planets repeat the cycle of their movements – but we cannot linger over the *Dream*. The cosmos represented in it, that of a finite series of concentric spheres revolving round the earth, remained dominant until the sixteenth century, even though Cicero's over-simplified version was known by experts even before his own day not to correspond to actual astronomical observation.

It will be seen that this pattern, while geocentric, accepts the sphericity of the earth, and it is perhaps worth insisting upon the fact that while in the Middle Ages uneducated people and occasional cranks sometimes supposed the earth to be flat, its sphericity was taken for granted by all educated folk.[11] It would be easy to demonstrate this from written texts – Bede, in order to make it clear that the earth was not simply a shield-like disc (*scutum*), expressly compares it to a ball with which boys play (*pila*)[12] – but it is perhaps most immediately apparent from the orb and cross, a familiar symbol of royalty in the West from the eleventh century onwards. Roman Emperors had often been shown holding a globe, symbolising the world they ruled, and in the sixth century Procopius, describing the great equestrian statue of Justinian in Constantinople, had drawn particular attention to the orb and cross the emperor held in his hand and explained its symbolism, the globe representing the world and the cross the divine authority by which he ruled it.[13] This is underlined by the old term *mound* which was used in England for the royal orb, for *mound* is no more than the French *monde* and its use is evidence of what men thought the world's shape to be.

How then, did the idea arise that medieval man believed the earth to be flat? Partly, no doubt, because there were some who held this view. The most curious world picture was that of a sixth-century Egyptian monk, Cosmas Indicopleustes, whose *Christian Topography*[14] was based on the literal interpretation of a selection of biblical texts. These showed, he believed, that the earth had the form of an oblong quadrilateral, sloping slightly from south to north – which is why the Nile, having to flow uphill, is a slower stream than the Tigris or Euphrates – and surmounted by an arched sky. A huge mountain in its northern part, by intercepting the rays of the sun, produced the distinction between day and night. This piece of eccentricity, never translated into Latin and totally without influence even in Byzantium, has never ceased to excite the derision of historians of science since its discovery and publication in the late seventeenth century. More important was the confusion of the idea of antipodes – persons living on the other side of the globe – with that of the shape of the world itself. The existence of antipodes was widely rejected by theologians,[15] since it seemed to follow from the teaching of the Bible (which made all men descend from Adam) and of classical geographers (who alleged that it was impossible to sail round the earth: if so, how could the antipodes 'down under' have got there?), and a few of the Church Fathers had gone on to reject the sphericity of the earth as well. The views of one of these, Lactantius,[16] a valuable if prejudiced historian of the early fourth century but a poor theologian, were paraded at length in Washington Irving's picturesque account of the commission which met in 1486 at Salamanca to examine Columbus's proposals,[17] and it is, one suspects, through this enormously successful work of the nineteenth century that the idea of a medieval 'flat earth' became the widely current legend that it is.

Cicero's picture of the cosmos in the *Dream of Scipio* was, quite legitimately, an oversimplified one. Since the spheres were crystalline and translucent the planets and stars could all be seen together, and each sphere had its own motion, circular in pattern but varying in speed, so that their movements would account for the changing features of the sky throughout the year. The spaces between the spheres were filled with a tenuous ether which allowed for the transmission of their influence. The pattern was fundamentally Aristotelian, based on a mixture of observation and reasoning. There were four kinds of matter, earth, water, air, and fire – we would still allow the first three if we changed their names to solid, liquid, and gas – each of which was thought to have its proper characteristics, the nature of earth being to sink and that of fire to rise, with water and air occupying intermediate positions between the two. The nature of earth to sink was indeed one of the main Aristotelian arguments for a geocentric universe, since it was there that all the heavy matter in the universe would tend to collect. The motion of inanimate objects, unless influenced by some external force, was uniform and either linear or circular, linear upwards or downwards for everything within the range of the earth, circular for the great wheeling pattern of the heavens and all that it contained. Circular motion was of a higher order than linear, for the circumferences of circles have neither beginnings nor ends. Pagan antiquity had treated the stars and planets as divinities, and it is often difficult, in the metaphorical terms of later writing, to know exactly what was believed of the 'intelligences' by which they were thought to be moved. (It should be remembered that Antiquity and the Middle Ages did not draw a hard and fast line between living and non-living matter: ores and rocks were usually regarded as alive, though with a 'life' many degrees inferior to that of fish and plants). Corruption and change were limited to the region extending as far as the moon, while beyond it the planets and stars moved in uniform circular patterns as they had done without change since the beginnings of recorded time. Comets and shooting stars were thought to be phenomena of the atmosphere, like clouds.

This splendid vision had unfortunately the major disadvantage of being wrong. The planets do not move round the earth, but round the sun; their motions are not circular, but elliptical; and their speed is not uniform, but varies according to their position on the ellipse. It is true that for the planets whose motions are most easily determined the deviation from perfect circularity is quite small, but already by Alexandrian times it was realised that observed positions were not always quite what would have been anticipated from the simple Aristotelian pattern. Various mathematical or physical devices were consequently introduced to 'save the phenomena'. The earth was not at the exact centre of the wheeling spheres, but eccentric, and the perturbations of the planets, the fact that they were sometimes ahead of and sometimes behind positions one would expect, were accounted for on the assumption that they moved in epicycles, revolving each in a small circle having its centre on the circumference of the greater circle it described in its motion round the earth or some point close to this. The resulting pattern, one of great mathematical refinement,[18] is that which in antiquity was summed up by Ptolemy and was vitally necessary for practising astronomers, even though its details might be a matter of indifference to ordinary folk.

If the structure of the universe was taken from Greece, knowledge of its origins was based upon the Bible, whose account of the six days of creation was the subject of innumerable commentaries from the Patristic age onwards. The works of the various days are familiar: light and darkness, called Day and Night, on the first day; the firmament known as Heaven, separating the waters above and below, on the second day; the dry land, the oceans, and the plants on the third day; the sun, the moon, and the stars, on the fourth day; the fish (including whales) and fowl on the fifth day; and the remaining animal kingdom, including man himself, made in God's image, on the sixth. There follows the story of the Garden of Eden, of 'Man's First Disobedience' and the acquisition of a knowledge of good and evil, and then the early history of mankind, of which the more sensational episodes are the stories of the Flood, of Noah's Ark, and of the Tower of Babel, till the historical parts of the Old Testament come finally to concentrate on the fortunes and misfortunes of the Chosen People.

This complex of traditions, going back to remote antiquity, included little on the structure of the universe or its shape, and the casual and poetic references found elsewhere in the Scriptures were for the most part never taken very literally. The creation story, however, seemed both detailed and authoritative, and was to occasion many difficulties, since it appeared in certain respects self-contradictory and in others ran counter to much that Greek tradition felt to be proved by reason. Further, while of only marginal importance in Jewish thought, which was prepared to accept it as poetry, it assumed an immense importance in Pauline Christianity because of the role ascribed to Christ in the doctrine of the Redemption, which presupposed the Fall of Man and the concept of Original Sin. 'For as in Adam all die, even so in Christ shall all be made alive.' In any case, the Biblical narrative was authoritative, while that of Greek science was not, at least in the same sense. *Quidquid in Sacra Scriptura continetur, verum est,* was in due course to become axiomatic; admittedly secular learning might sometimes teach otherwise, *sed in contrarium sufficit auctoritas Scripturae.*[19]

There is no need here to develop the theme of these contradictions, and show how they were either resolved or, more frequently, shuffled under the carpet. The weight attached to them varied from one commentator to another. Some were disturbed by the same kind of contradictions that might strike ourselves: the existence of night and day before the creation of the sun and moon, the creation of birds preceding that of the creeping things upon the earth. Others arose out of the philosophical preconceptions of the Greeks. If it is assumed that it is 'natural' for water to be heavier than air, it was difficult to explain how there could be waters above the firmament, even allowing for the fact that they might perform some valuable cooling function in the fiery outer reaches of the cosmos. Equally, to thinkers in the Greek tradition, there were difficulties in explaining how a timeless deity operated a creation in time, or how, if having made one world and found that it was good, he would not be bound to go on and create a plurality of worlds, since several 'goods' are better than one.

In any case, this Hellenic-Judaic-Christian cosmos was one that had to be populated, and here complications crept in that were not obviously required by the Creation story. Above man there were imagined various grades of superior beings, angels of all kinds, the cherubim and seraphim of scripture, the 'Thrones, Dominations, Virtues, Princedoms, Powers' of Milton, who found it necessary to subordinate their proper ordering to the needs of blank verse. These creatures of the ether in part grew out of Jewish folklore, where Jacob could wrestle with an angel and Abraham entertain them unawares, in part out of Neo-Platonic ideas, where God was too remote to occupy himself with the task of Creation and empowered a hierarchy of intermediate creatures to perform it, in part out of reasoning and analogy, for if the sea and earth and sky were inhabited by fish and quadrupeds and birds, how could the heavens be envisaged as lying useless with no living creatures to occupy them. (Kepler, after all, wondered whether comets might not be the whales of outer space.) What may be regarded as the accredited system was worked out by the fifth-century author of the *Celestial Hierarchy*, a mystical work which acquired a spurious authority through its authorship being attributed to the Dionysius the Areopagite who was one of St Paul's converts at Athens, till in the end Aquinas in his *Summa* found it necessary to devote fifteen *quaestiones* to a discussion of angelic qualities.[20]

Below the angels, in this geocentric universe, was man himself, ambiguously placed because while on the one hand he had been made in the image of God and was thought worthy that the Son of God should die for his salvation, on the other he had fallen from his high estate, and in consequence was himself exposed to every kind of misfortune, including death itself. For some theologians, and for perhaps most moralists, he was little more than a miserable sinner, a monster of wickedness; for humaner thinkers he was only a little lower than the angels, crowned with glory and honour, still exercising the rights of dominion over the fish of the sea, and the fowl of the air, and every living thing that walked upon the earth, as he had been promised in the book of Genesis. 'Who, then, can fully express the pre-eminence of so singular a creature? Man crosses the mighty deep, contemplates the range of the heavens, notes the motion, position, and size of the stars, and reaps a harvest both from land and sea . . . He foretells the future, rules everything, subdues everything, enjoys everything. He converses with angels, and with God himself.' So a Neo-Platonist turned Christian of the late fourth century, Nemesius of Emesa, in his *De Natura Hominis*,[21] a work which admittedly received wider authority than it would have otherwise had through erroneously being attributed to Gregory of Nyssa, and which, having been translated into Latin in the eleventh century, played some role in late medieval and Renaissance thought.

If the Christian cosmos of the Middle Ages and its inhabitants had a past, they also had a present, in which God's hand was seen to be constantly at work, and a future. These were things which did not bulk very large in Greek tradition. In Homer the gods interfere capriciously in human life; in fifth-century Athens, outside the highly moral world of the great tragedians, they often play little more than a formal role – 'Sacrifice a cock to Aesculapius'; in later Greek thought the relationship between the individual and the godhead bulk overwhelmingly large; but nowhere, or almost nowhere, is there any notion of a divine purpose controlling the course of history. This, on the other hand, was profoundly important in Jewish thought, where God is represented as caring for His chosen People, presiding over their history, punishing them all too frequently for disobedience and backsliding, and promising them in the fulness of time a Messiah. Christianity naturally took over the concept of a Divine guidance in history, and in the Apocalypse provided a lurid account of the Last Things, even if its horrors are followed by an eventual vision of the Holy City, the new Jerusalem, coming down from God out of Heaven, after the first heaven and the first earth have passed away. This is obviously high poetry, not scientific cosmology, but it is necessary to remember that the traditional picture of the universe was not just a matter of literary or poetic allusion; it was thought to be hard matter of fact. Early medieval cosmology looked back to that of antiquity, and even if it was mainly received on the basis of faith and tradition it was ultimately based on the reasoning of Aristotle and an accumulation of astronomical observations finally embodied in the great synthesis of Ptolemy. The writings of neither of these could be consulted at first hand, but enough was known to permit scholars to measure time, to compute planetary and stellar positions and eclipses, and to carry out the other duties expected of the astronomers and astrologers of the time.

With the twelfth and thirteenth centuries a new epoch began. Its most obvious feature was an enormous increase in the range of material at the disposal of Western scholars.[22] The chief surviving products of Greek astronomical thought not so far available to them were, from the practical point of view, Ptolemy's *Syntaxis*, an encyclopedia of astronomy which had been translated into Arabic by al-Ḥajjāj ibn Maṭar in 829/30 under the title of *Kitāb al-mijisti* (whence our *Almagest*) and from the theoretical point of view Aristotle's *De caelo* and parts of his *Meteorologica* and his *Metaphysics*. The *Almagest* and the *De caelo* were translated into Latin by Gerard of Cremona in 1175, the fourth book of the *Meteorologica* (which is not by Aristotle) directly from Greek by Aristippus of Catania in c. 1160 and the first three books from Arabic by Gerard of Cremona a little later, and the *Metaphysics* directly from Greek shortly after 1204. Although these early Aristotelian translations were in part or wholly superseded by those made in the mid-thirteenth century by William of Moerbeke, archbishop of Corinth, at the instance of St Thomas Aquinas, they had a very wide circulation in their day. The Arabic works translated were partly tables and astronomical data, partly commentaries on Aristotle or Ptolemy, and partly original works.

This is not the place for an account of Muslim astronomy, but one must at least mention the names of al-Khwarīzmī (Alcoarismi), whose astronomical tables were translated by Adelard of Bath in 1136 and were the most distinguished forerunners of the thirteenth century Alphonsine Tables of Toledo (1272), Māsh ā'allāh, the eighth century author of a celebrated treatise on astrology, the ninth century Baghdad geometer and astronomer Thābit ibn Qurra, and the eleventh century Spanish instrument maker and astronomer al-Zarkali (Arzachel), who wrote on the astrolabe. The commentary on the *De caelo* by Ibn Rushd (Averroës), translated by Michael Scot (*c.* 1120), was one of the seminal works in thirteenth century western thought, and Michael also translated (1217) the astronomical treatise of his older contemporary al-Biṭrūjī (Alpetragius), whose views were Aristotelian rather than Ptolemaic but who introduced the theory of impetus developed though not invented by John Philoponus in the sixth century. Later in the Middle Ages, under the Il-Khans of Persia, an important centre of observational astronomy was developed at Marāgha in Asia Minor under Nasīr ad-Dīn Tūsī, which made great improvements in instruments, though its influence on the astronomical thought of the west is still a matter of debate. The kind of way this writing could influence the west can be shown from a recent summary[23] of the questions discussed by Michael Scot's commentary, a work otherwise without importance, on Sacrobosco's *De Sphaera*:

'Whether the world is from eternity, whether it is one or many, whether it will sometime end, what its form or figure is, whether there is a ninth sphere, whether a round body is moved more readily than an angular body, whether the earth is habitable at the equator, how many elements there are, whether they are transmuted, whether fire is hot in its own sphere, whether water surrounds the earth, whether the ethereal region is moved about the elementary, whether the heavens are moved by one mover, whether their movement is natural or violent or voluntary, whether the heavens are an animated body, whether they are moved faster than the stars, and whether uniformly.'

This material made more and more apparent the defects of the Ptolemaic system, so that it was natural for scholars to try and find a better model, which would dispense with the eccentrics and epicycles which were needed to make it work but which at the same time went far towards rendering it incredible. It is in this period that 'the Sphear' which Milton mocked,

> With Centric and Eccentric scribl'd o're,
> Cycle and Epicycle, Orb in Orb

really took shape. Averroës, the greatest scholar of the Muslim world, recalls touchingly how in his youth he had hoped to find some alternative, and how in his old age, despairing of success, he none the less put his ambition on record in the hope of encouraging other scholars, one of whom might have the good fortune denied to him.[24] Again and again the Schoolmen might discuss it, in commentaries on Book II of the *De caelo* or Book XII of the *Metaphysica*, but computational requirements always outweighed its flagrant disaccord with the assumptions of Aristotelian physics. As long as men's minds were bound by the notion of a geocentric universe, and of uniform and circular movements in the heavens, there was no escape. The possibility of extending to other planets the view traditionally ascribed to Heraclides of Pontus and passed on to the Middle Ages by Chalcidius and Martianus Capella, that Mercury and Venus, the planets 'inferior' to the sun, revolved round it and not round the earth, went practically unconsidered. Nicole Oresme's discussion of the possible rotation of the earth shows clearly how an intelligence of the highest order might remain limited by one of those conceptual necessities which can for so long block the road to scientific progress. He was quite capable of seeing and setting out the flaws in the arguments for believing that the earth must be stable, but it is no more than an intellectual exercise, and ends with the formal statement that 'everyone holds, and I think myself, that the heavens do move and not the earth, since "God established the round world (*orbem terrae* in the Vulgate) that it shall not be moved" (Ps. 92.2 in the A.V.), despite the arguments to the contrary.'[25] Scholars have tried to furnish Copernicus with precursors – Duhem's candidate was in fact Oresme, and in recent years al-Ṭūsī and Ibn al-Shāṭir, as representatives of the Marāgha school, have been billed for the same post[26] – but in the last resort no real breach was made with the old system.

Besides these problems of the sky itself, there were others which are not generally thought of as cosmological but whose understanding depends on our picture of the structure of the universe. The tides were known to be affected by the moon – *De concordia maris et lunae* is the title of a chapter in Bede's *De temporum ratione* – and their explanation raised such problems as the nature of their motion (was the water simply sucked upwards, being rarified in the process and sinking backward when the moon passed?) and how was it effected (by the moon-beams, and how did they operate, for the moon was far away and it was an established maxim that 'matter cannot act where it is not'?)[27] Rainbows were a perennial subject of speculation, and raised problems of reflection, refraction, the nature of light, and the theory of vision.[28] Shooting stars and comets were be-lieved to belong to the realm of things beneath the moon, since their disappearance did not diminish the number of the stars seen in the sky; what were they, and what function did they have? *Isti mirant stella(m)* is one of the captions in the Bayeux Tapestry, and the occurrence of comets was so rare that they might fairly be presumed to presage mighty events, as indeed they were held to do on that occasion.[29]

At a lower level, all could expect to be influenced by the day-to-day movements of the stars and planets. Astrology and cosmology had been related since very early times, and astro-logy had flourished in the Hellenistic period and under the Roman Empire. The Stoic doctrine of sympathy held that nothing could happen in the universe without it affecting, in however slight a manner, all other things, and the importance attributed to man, the deep-seated feeling that all things were ultimately created for his benefit or instruction, made it natural to assume that celestial bodies must influence his fate.[30] If the sun could affect the seasons and the moon determine the tides, why should not the conjunctions of the stars and planets determine the fortunes of states and the lives of individuals, particularly those of persons of rank whose luxurious diets, some thinkers suggested, rendered them less resistant to stellar influences, especially maleficent ones, than those of humbler extraction and tougher fibre. So, on the ancient assumption of all-pervading celestial influences, there had been gradually erected an extraordinary apparatus of astrological doctrine, isolating the influences proper to every hour of the day, every part of the body, every country of the world, prescribing how they should be observed in sowing fields and undertaking travel, in administering drugs and bleeding patients, noting the mysterious correspondence between the seven planets and the seven metals or the seven orifices of the body, providing rules for the casting of horoscopes based on the position of the constellations at the very moment of a man's birth or even – much more difficult to ascertain – his conception. The main stimulus to observation in the field of what was called 'natural' astrology (i.e. our 'astronomy') was not so much disinterested curiosity as the need to provide practitioners of 'judicial' astrology (i.e. the prediction of events on the basis of the position of the stars) with one of the necessary tools of their trade.

It was partly this interest in astrology that accounts for the extraordinary popularisation of astronomy in the late Middle Ages and the sixteenth century. Perhaps its first symptom was the immense success of the small handbook *De sphaera* of John of Sacrobosco (of Halifax or Holywood),[31] a work of the mid-thirteenth century which covers the whole field of cosmology. Although entirely derivative, its compactness – it runs to under forty pages of print – led to its being treated as a convenient school manual, and long after the invention of printing it continued to appear in Germany and Spain, either in Latin or in the vernacular, though its contents were by that time completely out of date. It is no less apparent in the knowledge of astronomy and cosmology displayed by Dante and Chaucer, the two greatest poets of the Middle Ages. That shown in Chaucer's early poems may have been no more than what might be expected of an ordinary educated person of his time, but he was a man of strong scientific interests and the astronomical allusions in his later work, together with his treatise on the astrolabe, show that he subsequently made a hobby of the subject and became fully competent in it.[32] Dante's knowledge was also exceptional,[33] though it sometimes gives the impression of being book-learning and its introduction rather contrived, but he has the distinction of having added to the customary picture of the universe in his detailed presentation of the geography of hell, which had previously been vaguely located beneath the earth but whose precise divisions no one had hitherto attempted to map out. Another poet, John Gower, devotes nearly a thousand lines in his *Confessio Amantis* to Alexander's lessons in astronomy at the hands of Aristotle. It has been unkindly said, however, that as a piece of instruction this shows 'an ineptitude remarkable even in the annals of pedagogy', and Gower seems to have got up the subject for the occasion without making any serious effort to understand it.[34]

During the Renaissance the hold of the traditional system was at first strengthened as a result of the classical revival, the new elements in which were the emigration of Byzantine scholars with their manuscripts to the West and the gradual spread of a knowledge of Greek. Fresh works in consequence came to light, and these could now be studied in their original language, not in frequently indifferent translations. The current veneration for antiquity also conferred on classical authors the reputation for infallibility formerly reserved for Holy Writ. Further, while one stream of Renaissance thought can be regarded as basically humanist and rational, many scholars succumbed to an attack of Neo-Platonism and an interest in the Hermetic, magical, and astrological traditions of late anti-

quity. The Hermetic Corpus, in company with Plotinus and
Iamblichus, was translated and commented upon by Marsilio
Ficino and assimilated by Pico della Mirandola to Jewish
mystical traditions embodied in the Kabbalah, and thence-
forward a series of distinguished if muddled thinkers – Trithe-
mius, Cornelius Agrippa, Paracelsus, Bruno, Campanella, John
Dee – devoted themselves with varying degrees of success to
an astonishing mixture of mathematics and magic, science and
nonsense. Conversations with angels became a major ambition
of scholars, and astrological prognostications reached a pitch
of abstruseness that they had never known in the Middle Ages.
Yet it was at this point, where the old cosmology reached its
final peak of complexity, that its days were numbered.

Our interpretation of the fall of the old cosmology depends
very largely upon what we regard as its essence. We can look
upon it as a basically authoritarian system, defended on the
ground of its essential rationality but in fact resting on the
prestige of a part pagan, part Biblical antiquity. From this
point of view, the discovery of serious error in a world picture
so intimately held together by logical bonds would imply that,
sooner or later, the whole scheme would come apart. Or we
can emphasise its geocentric character, and see its end in the
general acceptance of a heliocentric universe as a more satisfactory alternative. Or we can regard it as essentially anthropocentric, in which case the crucial events are likely to be the
abandonment of the Creation story of Genesis and the acceptance of the doctrine of evolution, integrating man into the
general pattern of living creatures, with all that this implied for
the traditional Christian doctrines of the Fall and of Redemption.

Most scholars tend to adopt one or the other of the two latter
standpoints, though in each case the story they have to tell is a
complex one. In the disappearance of the geocentric universe,
the major events are well known:[35] Copernicus and the *De
Revolutionibus Orbium Coelestium* of 1543, with its ambiguous
preface by Osiander suggesting that the hypothesis was a
mathematical device, not a true model; Tycho Brahe and his

De Nova Stella (1573) describing the nova in Cassiopeia of November 1572 which, in a period of accurate and trustworthy astronomical observations, showed that the heavens were not unchanging and incorruptible, as the whole Aristotelian tradition asserted; Kepler and the theory of elliptical orbits, propounded in his *Astronomia Nova* of 1609, which by attributing to the planets a non-circular motion broke further away from the Greek tradition and its cherished belief of perfection in the skies; Galileo and his telescope, demonstrating the nature of the Milky Way and more complexities or imperfections in the solar system – the moons of Jupiter, the spots upon the sun – and his resort to the vernacular in support of the Copernican system; Newton and his laws and the universal theory of gravitation. Anthropocentrism has had a longer life, and in a sense it is still with us, though in its medieval form it can no longer be maintained. The crucial dates in its downfall are usually taken as 1859 and 1871, those of the publications of Darwin's *On the Origin of Species* and *The Descent of Man* respectively, but at least as important are 1735, when Linnaeus published the first edition of his *Systema Naturae*, firmly classing man with the apes and monkeys amongst the primates, and 1846, when Boucher de Perthes, on the basis of flint implements found in the gravels of the Somme valley, first obtained clear proof of the antiquity of man.[36] The shock of Linnaeus' classification can be judged by the reaction of a younger contemporary, the English naturalist Thomas Pennant, who in 1767 wrote to dissuade Joseph Banks from going to study at Uppsala and in 1781, in his *History of Quadrupeds*, stated his prejudices firmly. 'My vanity will not suffer me to rank mankind with Apes, Monkies, Maucaucos and Bats.'[37]

These two approaches are defensible, but a case can be made out for the view that the geographical discoveries of the fifteenth and sixteenth centuries were more crucial still. One is familiar with the problems raised for theologians, as well as for political thinkers by the discovery of America.[38] These problems are sometimes treated as purely human – how did the native population originate? might the Indians be legitimately conquered or enslaved? – but those raised by the American fauna were just as disconcerting: what were its origins, why is much of it not represented in the Old World, and, in the light of this, could its ancestors really be held to have emerged from the Ark? But the earlier crossing of the Equator, and the Portuguese voyages to India, marked in a more real sense the birth of something new, since they cast doubt upon the infallibility of the classics. There is a deeply revealing passage in an oddly entitled tract, *Esmeraldo De Situ Orbis*, written in about 1505 by a Portuguese mariner, Duarte Pacheco Pereira, describing the sea route to India. The author justifiably exults in the great achievements of his own nation, and goes out of his way to insist on how they had disproved the assertions of the ancient geographers. 'Ptolemy', he observed,[39]

'in his portrayal of the ancient tables of cosmography writes that the Indian Ocean is like a lake, far removed from our western Ocean which passes by southern Ethiopia; and that the voyage was so long as to be impossible and that there were many sirens and great fishes and dangerous animals which made navigation impossible. Both Pomponius Mela . . . and Master John Sacrobosco, an English writer skilled in the art of astronomy . . . said that the country in the Equator was uninhabitable owing to the great heat of the sun, and since it was uninhabitable for this reason it could not admit of navigation. But all this is false (*o que tudo isto he falso*), and we have reason to wonder that such excellent authors as these, and also Pliny and other writers who averred this, should have fallen into so great an error . . . Since experience is the mother of knowledge, it has taught us the absolute truth (*e como quer a experiencia he madre das cousas, por ella soubemos rradicalmente a verdade*); for our Emperor Manuel, being a man of enterprise and great honour, sent out Vasco de Gama . . . to discover and explore those seas and lands concerning which the ancients had filled us with such fear and dread; and after great difficulty, he found the opposite of what most ancient writers had said.'

This passage sounds a new note in the history of Western

thought. Pacheco's reaction, with its exultant overtones, was no doubt premature. Guicciardini, writing three decades later, while outspoken in his appreciation of the Portuguese discoveries and those of Columbus, realised that more was in peril than the reputation of the classics. 'Not only has this navigation confounded many things asserted by former writers about terrestrial things; it has also given some anxiety to interpreters of Holy Scripture.'[40] The implications of Lactantius' errors on the sphericity of the earth were in fact underlined by Copernicus in the dedication of his *De revolutionibus* to Pope Paul III. As time wore on the authority of the ancients, pagan, Jewish and Christian alike, was more and more going to be set aside, and Experience to be regarded as the Mother of Knowledge. On the size of the world, and the navigability of the oceans, authority had been shown to be wrong. The discarding of the old image had begun, and the new heavens and the new earth of modern cosmology, linked curiously to the old by an equal faith in the existence of a mathematical structure of reality behind external appearances, had started to emerge.

Notes

1. This paper is printed virtually as it was delivered in lecture form, apart from slight verbal alterations, a few additions here and there, and the inclusion of some pages on late medieval cosmology which had to be omitted from the spoken version owing to lack of time. I am grateful to Professor C. N. L. Brooke and Mr G. E. R. Lloyd for having read it through and suggested clarifications and improvements. An account of the basic material will be found in two major reference works: G. Sarton, *Introduction to the History of Science*, Vols I–III (ii) (Baltimore, 1927–48), and P. Duhem, *Le système du monde: Histoire des doctrines cosmologiques de Platon à Copernic*, 10 vols (reprinted Paris, 1954–9; unfinished, vols. vi–x added posthumously in 1954–9 to the five volumes of the 1913–17 edition, extending to the end of the fifteenth century). The best general account is still that of J. L. E. Dreyer, *A History of Astronomy from Thales to Kepler* (New York, 1953); but since this is an unrevised reprint

(although with supplementary bibliography) of a work originally published under a slightly different title in 1906, some of it is now out of date. For the early Middle Ages, see the relevant sections in W. H. Stahl, *Roman Science: Origins, Development, and Influence to the Later Middle Ages* (Madison, Wis., 1962) and further articles referred to below; also, although now itself somewhat outdated, A. van de Vyver, 'L'évolution scientifique du haut moyen age', *Archeion*, xix (1937), 12–20. J. L. E. Dreyer's 'Medieval astronomy' in *Studies in the History and Method of Science*, C. Singer (ed.), II (Oxford, 1921), 102–20, is a general essay based on Duhem. There is a good account in the relevant sections of A. C. Crombie, *Augustine to Galileo* (1952). Those in Vol. I of *La science antique et médievale*, R. Taton (ed.) (Paris, 1957) are informative but very confused in their presentation. The French text is preferable to the English translation (1963).

2. Walter Lippmann, *A Preface to Morals* (London, 1929), 68–9.

3. This contrast, popularised through the writings of Mircea Eliade (*Le mythe de d'éternal retour*, revised ed. Paris, 1969), has, however, been much exaggerated. There is a good demolition job by A. Momigliano, 'Time in ancient historiography', *History and Theory*, Beiheft 6 (1966), 1–23.

4. See the evaluation in R. Lenoble, *Esquisse d'une histoire de l'idée de Nature* (Paris, 1969), 137 ff.

5. J. Wrobel's edition (Leipzig, 1876) is now superseded by that in the *Corpus Platonicum Medii Aevii* by P. J. Jensen and J. H. Waszink, *Plato: Timaeus, a Calcidio translatus commentarioque instructus*, Plato Latinus, 3 (London-Leiden, 1962). There is a useful analysis of the cosmological sections in W. H. Stahl, 'Dominant traditions in early medieval Latin science', *Isis*, l (1959), 121–3. Also relevant is J. C. M. van Winden, *Chalcidius on matter: his doctrine and sources* (Leiden, 1959).

6. This reservation is necessary because in the earlier period many subsequently important works had not yet been sufficiently copied to become widely accessible. Bede himself knew Macrobius only in excerpts, and Martianus Capella and the Latin Chalcidius not at all.

7. In Books iii, 24–71 (Astronomy), xiii, and xiv of his *Etymologiae*, W. M. Lindsay (ed.), 2 vols (Oxford, 1911) a much used collection of factual information, eschewing explanations (other than fanciful etymologies) and figures. Though it served as a kind of dictionary-encyclopaedia its cosmographical sections are quite different in character from the extended expositions of Macrobius and Martianus Capella. Cf. E. Brehaut, *An Encyclopedist of the Dark Ages: Isidore of Seville* (New York, 1912), and J. Fontaine, *Isidore de Séville et la culture classique dans l'Espagne wisigothique*, 2 vols (Paris, 1959).

8. *In Somnium Scipionis*, F. Eyssenhardt (ed.) (Leipzig, 1893). English translation, with excellent introduction and commentary, by W. H. Stahl:

Macrobius: Commentary on the Dream of Scipio (New York, 1952). See also Stahl, *Dominant Traditions*, 111–21. The recent redating of Macrobius' writings to the 420s is more important for his *Saturnalia* than for his commentary on the *Somnium Scipionis* (A. Cameron, 'The Date and Identity of Macrobius', *Journ. Roman Studies*, lvi [1966], 25–38).

9. *De nuptiis Mercurii et Philogiae*, A. Dick and J. Préaux (eds) (Stuttgart, 1969: a corrected reprint of the 1925 Teubner text, by Dick alone). There is no English translation, but the cosmological sections, which in the Middle Ages often circulated as a separate work, are very fully analysed by Stahl, *Dominant Traditions*, pp. 98–111. The MSS of his work, which form important evidence for his influence, have in recent years been studied by C. Leonardi, while a number of important ninth-century commentaries have come to light. For a guide to the literature, see W. H. Stahl, 'To a Better Understanding of Martianus Capella', *Speculum*, xl (1965), 102–15.

10. The translation is that of Stahl, pp. 72–5, *Macrobius: Commentary on the Dream of Scipio* (New York, 1952).

11. Cf. the informative but – perhaps appropriately – rather superficial sketch by C. W. Jones, 'The Flat Earth', *Thought*, ix (1934), 296–307, and the views of various patristic writers in Dreyer, *A History of Astronomy*, 207 ff.

12. *De temporum ratione*, c. 32, in his *Opera*, J. A. Giles (ed.) (1843), 210.

13. Procopius, *De aedificiis*, i.2.11. The ramifications of the symbol in European art and thought are abundantly illustrated in P. E. Schramm, *Sphaira, Globus, Reichsapfel* (Stuttgart, 1958).

14. Trans. by J. W. McCrindle, *The Christian Topography of Cosmas, an Egyptian monk*, Hakluyt Soc., xcviii (1897). Cf. W. Wolska, *La Topographie Chrétienne de Cosmas Indicopleustes: théologie et science au VIe siècle* (Paris, 1962), and, especially for its treatment of the illustrations, her new edition (not yet complete) of the text (Paris, 1968, 1970), which is accompanied by a French translation.

15. For example, St Augustine, *De civitate Dei*, xvi. 9.

16. Lactantius, *Divinae Institutiones*, iii. 24. As a trained rhetorician he pours scorn on the whole idea – can anyone believe in people walking upside down, in trees growing with their branches downwards, in it raining and snowing upwards, etc?

17. *Life and Voyages of Christopher Columbus*, 2nd revised ed. (1849), Book I, Ch. 3. There are no records of the proceedings at all, and the account is pure fantasy, but the introduction of Lactantius was perhaps suggested by Copernicus' later reference to this writer's errors (above, p. 252). Fernando Colombo mentions the doubts of Augustine on the antipodes (*Le historie della vita e dei fatti di Cristoforo Colombo*, c. 12; R. Caddeo (ed.) [Milan, 1930] I, 107) which is the source of the *duda Sant*

Agustin which Bartolomé de las Casas represents as the refrain of the opposition (*Historia de las Indias*, i. 29; L. Hanke (ed.) [Mexico City, Buenos Aires, 1951], I, 157).

18. The versatility and elegance of a pattern of epicycles in accounting for elliptical motions is well set out by N. R. Hanson, 'The Mathematical Power of Epicyclical Astronomy', *Isis*, li (1960), 150–8. The standard expositions of the Ptolemaic system are in need of some revision. See Appendix 2 to the second edition of O. Neugebauer, *The Exact Sciences in Antiquity* (Providence, R.I., 1957).

19. Aquinas, *Summa Theologiae*, Pars I, Quaest. 69, Art. 1 (against objections to the Biblical account of the second day of creation), Quaest. 70, Art. 1, 2 (those of the third day), etc. Augustine, in the passage cited above (p. 254, note 15), rejecting the antipodes, had asserted categorically that the Scriptures could in no fashion (*nullo modo*) err.

20. *Summa Theologiae*, Pars I, Quaestiones 50–64. St Thomas' views on creation follow in Quaest. 65–74, but his general cosmological views are set out in his commentary on Aristotle's *De Caelo*, which uses Moerbeke's translation of this work and Simplicius' commentary on it. Aquinas' own commentary, composed at Naples in 1272–3 and one of his last works, has been described as 'the high water-mark of St Thomas' expository skill'.

21. W. Telfer, *Cyril of Jerusalem and Nemesius of Emesa* (1955), 255. The work, for which Telfer provides an excellent introduction and commentary, was apparently composed in draft before Nemesius' conversion and subsequently only in part re-written. See also Telfer's article 'The Birth of Christian Anthropology', *J. Theol. Studies*, N.S., xiii (1962), 347–54.

22. The best guide to translations of this period are the prefaces to Vols II and III of Sarton's *Introduction to the History of Science* (esp. II [i], 20–1, 113–16, 167–75). For those from Arabic (but excluding those of Aristotelian commentators), see F. J. Carmody, *Arabic astronomical and astrological science in Latin translation: a critical bibliography* (Berkeley, 1956). Earlier ones are dealt with by A. van de Vyver, 'Les plus anciennes traductions latines médiévales (Xᵉ-Xᴵ siècles) de traités d'astronomie et d'astrologie', *Osiris*, i (1936), 658–91. For translations of Aristotle and his commentators see L. Minio-Paluello in *Dictionary of Scientific Biography*, I (New York, 1970), 270–5. These are now being published in two series, the *Corpus philosophorum medii aevi: Aristoteles Latinus* (Bruges-Paris, 1952–) and the *Corpus Latinum commentariorum in Aristotelem Graecorum* (Louvain, 1957–).

23. L. Thorndike, *Michael Scot* (London, 1965), 37.

24. '*In iuventute autem mea speravi, ut haec perscrutatio completetur per me: in senectute autem iam despero, sed forte iste sermo inducet aliquem ad perscrutandum de hoc*', in his commentary on Book XII of Aristotle's

Metaphysics: cited by C. Kren in her discussion of the unsuccessful attempt at simplification made by a fourteenth century schoolman, 'Homocentric Astronomy in the Latin West: the *De reprobatione ecentricorum et epiciclorum* of Henry of Hesse', *Isis*, lix (1968), 269–81; p. 269, note 1.

25. Oresme's commentary on the *De caelo* of Aristotle, entitled *Le livre du ciel et du monde*, is edited and translated by A. D. Menut and A. J. Denomy (Madison, Wis., 1968); the passage cited is on p. 536. It is basically the argument of his older contemporary Jean Buridan, who in Question 22 of his commentary on Book II of Aristotle's *De caelo* had maintained the diurnal rotation of the earth, *Quaestiones super libros quattuor de caelo et mundo*, ed. E. A. Moody (Cambridge, Mass., 1942), 226–33; cf. J. Bulliot, 'Jean Buridan et le mouvement de la terre', *Revue de philosophie*, xxv (1914), 5–24. In the fifteenth century Nicholas of Cusa, on grounds of the relativity of motion, also argued for a moving earth, maintaining also that the whole world picture was a pure concept, since space could not be thought of as having boundaries or the universe as having a centre. See Book II, Chs 11 and 12, of his *De docta ignorantia*, trans. by G. Heron, *Nicolas Cusanus: Of learned ignorance* (London, 1954), 107–18.

26. P. Duhem, 'Un précurseur français de Copernic: Nicole Oresme', *Revue générale des sciences pures et appliquées*, xx (1909), 866–73. For the Marāgha school see E. S. Kennedy, 'Late medieval planetary theory', *Isis*, lvii (1966), 365–78, and earlier articles there cited; for claims on its behalf, Seyyed Hossein Nasr, *Science and civilization in Islam* (Cambridge, Mass., 1968), 172–4, commented on sceptically by B. S. Eastwood, in *Speculum*, xlv (1970), 152.

27. Cf. R. C. Dales, 'The text of Robert Grosseteste's *Questio de fluxu et refluxu maris* with an English translation', *Isis*, lvii (1966), 455–74, partly reflecting an explanation put forward by al-Biṭrūjī, on the basis of his astronomical theories.

28. C. B. Boyer, *The rainbow: from myth to mathematics* (New York, 1959), and D. C. Lindberg, 'Roger Bacon's theory of the rainbow: progress or regress?', *Isis*, lvii (1966), 235–48.

29. L. Thorndike, *Latin treatises on comets between 1238 and 1368 A.D.* (Chicago, 1950), a series of texts (in Latin) with explanatory introductions, and his article, 'Peter of Monte Alcino's Treatise on Comets', *Isis*, xl (1949), 350–1.

30. Lynn Thorndike's *History of Magic and Experimental Science* (8 vols., New York, 1923–58) is the chief mine of information. Cf. also M. Graubard, *Astrology and Alchemy: Two Fossil Sciences* (New York, 1953); Thorndike's argument that the assumptions of astrology, in relating the heavenly bodies to earthly happenings, should be regarded as a 'natural law', erroneous indeed but not in their essence different from

the Newtonian theory of gravitation ('The True Place of Astrology in the History of Science', *Isis*, xlvi [1955], 273–78); and Graubard's reply ('Astrology's Demise and its Bearing on the Decline and Death of Beliefs', *Osiris*, xiii [1958], 210–61). The study of T. O. Wedel, *The Mediaeval Attitude toward Astrology* (New Haven, 1920), is mainly concerned with English literature and so more limited than the title suggests. D. C. Allen, *The Star-crossed Renaissance* (Durham, N.C., 1941) is excellent.

31. Modern edition and translation by L. Thorndike, *The Sphere of Sacrobosco and its Commentators* (Chicago, 1949). On its long period of influence, cf. Sarton, *Introduction*, II, 617–19. Second only to Sacrobosco in popularity – over 200 MSS are known – was a late thirteenth-century work of unknown authorship which gave a much better account of planetary theory. See O. Pedersen, 'The *Theorica Planetarum* – Literature of the Middle Ages', *Classica et Mediaevaila*, xxiii (1962), 225–32.

32. Chaucer's treatise on the astrolabe is in W. W. Skeat, *Complete Works of Geoffrey Chaucer* (Oxford, 1894), III, 175–232, and an account of how to construct an instrument for computing planetary positions, apparently in his actual handwriting, has been published by D. J. Price, *The Equatorie of the Planets* (Cambridge, 1955). Chauncey Wood's *Chaucer and the Country of the Stars* (Princeton, 1970) is a mine of information. Cf. also H. M. Smyser, 'A View of Chaucer's Astronomy', *Speculum*, xlv (1970), 359–73.

33. Standard work by M. A. Orr (Mrs John Evershed), *Dante and the early astronomers*, 2nd ed (London, 1956), the revised edition of a book first published in 1913 by an author who was both an Italian scholar and a competent astronomer, her husband having been on the staff of Kodaikanal Observatory (South India).

34. Gower, *Confessio Amantis*, Book VII, lines 631–1506, in *The Complete Works*, ed. G. C. Macaulay, III (Oxford, 1901), 250–73. The comment is from Smyser's article just cited, p. 361. Cf. also G. G. Fox, *The Mediaeval Sciences in the Works of John Gower* (Princeton, 1931), Chs 3 and 4.

35. There is no point in attempting a bibliography. The best general account is that of T. S. Kuhn, *The Copernican Revolution* (Cambridge, Mass., 1957), and for the impact of the 'new' astronomy in this country F. R. Johnson, *Astronomical Thought in Renaissance England* (Baltimore, 1937).

36. These discoveries were at first regarded as fraudulent, and only fully accepted as a result of visits to the site by Joseph Prestwich, John Evans, and other members of the Geological Society in 1859. Their report is the standard vindication, but there is a little known account, based on Evans' diary, in Joan Evans, *Time and Chance* (1943), 100–4.

37. W. Blunt, *The Compleat Naturalist. A Life of Linnaeus* (1971), 154.

R

38. J. H. Elliott, *The Old World and the New, 1492–1650* (Cambridge, 1970), especially Ch. 2; L. Hanke, *Aristotle and the American Indian* (Chicago, 1959); and S. Jarcho, 'Origin of the American Indian as suggested by Fray Joseph de Acosta (1589)', *Isis*, 1 (1959), 430–8 (especially good on fauna).

39. D. Pacheco Pereira, *Esmeraldo de situ orbis*, iv. 1; translated by G. H. T. Kimble, Hakluyt Soc., 2nd ser., lxxix (London, 1936), 164–5. I cite Kimble's translation.

40. F. Guicciardini, *Storia d'Italia*, vi. 9, C. Panigada (ed.) (Bari, 1929), II. 132). I owe this reference to Professor Elliott. What especially disconcerted Guicciardini was the vast number of non-Christians there were now found to be, since Ps. xlviii. 10 had been comfortably interpreted as implying that most of the world had been made acquainted with the Gospel.

Index